Soft Robotics in Rehabilitation

Soft Robotics in Rehabilitation

Edited by

AMIR JAFARI
Advanced Robotic Manipulators ARM Lab, Department of Mechanical Engineering, College of Engineering, University of Texas at San Antoni UTSA, San Antonio, TX, United States

NAFISEH EBRAHIMI
Advanced Robotic Manipulators ARM Lab, Department of Mechanical Engineering, College of Engineering, University of Texas at San Antoni UTSA, San Antonio, TX, United States

Academic Press is an imprint of Elsevier
125 London Wall, London EC2Y 5AS, United Kingdom
525 B Street, Suite 1650, San Diego, CA 92101, United States
50 Hampshire Street, 5th Floor, Cambridge, MA 02139, United States
The Boulevard, Langford Lane, Kidlington, Oxford OX5 1GB, United Kingdom

Copyright © 2021 Elsevier Inc. All rights reserved.

No part of this publication may be reproduced or transmitted in any form or by any means, electronic or mechanical, including photocopying, recording, or any information storage and retrieval system, without permission in writing from the publisher. Details on how to seek permission, further information about the Publisher's permissions policies and our arrangements with organizations such as the Copyright Clearance Center and the Copyright Licensing Agency, can be found at our website: www.elsevier.com/permissions.

This book and the individual contributions contained in it are protected under copyright by the Publisher (other than as may be noted herein).

Notices
Knowledge and best practice in this field are constantly changing. As new research and experience broaden our understanding, changes in research methods, professional practices, or medical treatment may become necessary.

Practitioners and researchers must always rely on their own experience and knowledge in evaluating and using any information, methods, compounds, or experiments described herein. In using such information or methods they should be mindful of their own safety and the safety of others, including parties for whom they have a professional responsibility.

To the fullest extent of the law, neither the Publisher nor the authors, contributors, or editors, assume any liability for any injury and/or damage to persons or property as a matter of products liability, negligence or otherwise, or from any use or operation of any methods, products, instructions, or ideas contained in the material herein.

British Library Cataloguing-in-Publication Data
A catalogue record for this book is available from the British Library

Library of Congress Cataloging-in-Publication Data
A catalog record for this book is available from the Library of Congress

ISBN: 978-0-12-818538-4

For Information on all Academic Press publications
visit our website at https://www.elsevier.com/books-and-journals

Publisher: Mara Conner
Acquisitions Editor: Sonnini R. Yura
Editorial Project Manager: Emily Thomson
Production Project Manager: Poulouse Joseph
Cover Designer: Matthew Limbert

Typeset by MPS Limited, Chennai, India

Contents

List of contributors ix
Introduction and acknowledgments xi

1. **High-performance soft wearable robots for human augmentation and gait rehabilitation**   1
   Antonio Di Lallo, Shuangyue Yu, Tzu-Hao Huang, Thomas C. Bulea and Hao Su

   1.1 Introduction   1
   1.2 Actuation technologies for physical human–robot interaction   2
   1.3 Applications to wearable robots   7
   1.4 Discussion   34
   Acknowledgments   35
   References   35

2. **Development of different types of ionic polymer metal composite-based soft actuators for robotics and biomimetic applications**   39
   Ravi Kant Jain

   2.1 Introduction   39
   2.2 Literature survey on IPMC as actuators and sensors and its applications   40
   2.3 Development of IPMC base soft actuator by different approaches   44
   2.4 Results and discussions   50
   2.5 Development of robotic system using different types of IPMC actuators   80
   2.6 Conclusion   82
   Acknowledgment   83
   References   83

3. **Soft actuators and their potential applications in rehabilitative devices**   89
   Alexandrea Washington, Justin Neubauer and Kwang J. Kim

   3.1 Introduction   89
   3.2 Overview of soft robotic actuators   91
   3.3 Applications of soft robotic actuators   101
   3.4 Conclusions   106
   Acknowledgments   107
   References   107

4. An optimized soft actuator based on the interaction between an electromagnetic coil and a permanent magnet — 111
Nafiseh Ebrahimi, Paul Schimpf and Amir Jafari

    4.1 Introduction — 111
    4.2 Solenoid magnetic field and force calculation — 113
    4.3 Solenoid geometry design optimization — 120
    4.4 Manufacturing aspects and limitations — 125
    4.5 The influence of solenoid section deformation on the magnetic field and force — 128
    4.6 Discussion and conclusion — 131
    Acknowledgment — 132
    References — 132

5. Cable-driven systems for robotic rehabilitation — 135
Rand Hidayah, Tatiana Luna and Sunil Agrawal

    5.1 Introduction — 135
    5.2 Cable-driven leg exoskeleton for gait rehabilitation — 149
    5.3 A perturbation study using Robotic Upright Stand Trainer — 156
    5.4 Conclusion — 159
    Acknowledgments — 159
    References — 159

6. XoSoft: design of a novel soft modular exoskeleton — 165
Jesús Ortiz, Christian Di Natali and Darwin G. Caldwell

    6.1 Introduction — 165
    6.2 User-centered design — 166
    6.3 Requirements — 169
    6.4 Actuation principle — 172
    6.5 Sensing and control — 175
    6.6 Prototypes — 179
    6.7 Testing and validation — 184
    6.8 Conclusions and future works — 195
    References — 196

7. TwAS: treadmill with adjustable surface stiffness — 199
Amir Jafari and Nafiseh Ebrahimi

    7.1 Introduction — 199
    7.2 Actuator with Adjustable Stiffness mechanism — 201

7.3 Experimental results for stiffness adjustment of Treadmill with Adjustable Stiffness 224
References 238

8. An artificial skeletal muscle for use in pediatric rehabilitation robotics **241**
Ahad Behboodi, James F. Alesi and Samuel C.K. Lee

8.1 Introduction 241
8.2 Method 245
8.3 Results 246
8.4 Discussion 253
8.5 Conclusion 256
References 256

Index 259

List of contributors

Sunil Agrawal
Robotics and Rehabilitation Laboratory, Mechanical Engineering Department, Columbia University, New York, NY, United States

James F. Alesi
Biomechanics and Movement Science Program, University of Delaware, Newark, DE, United States

Ahad Behboodi
Biomechanics and Movement Science Program, University of Delaware, Newark, DE, United States; Department of Physical Therapy, University of Delaware, Newark, DE, United States

Thomas C. Bulea
Functional and Applied Biomechanics Section, Rehabilitation Medicine Department, Clinical Center, National Institutes of Health, Bethesda, MD, United States

Darwin G. Caldwell
Department of Advanced Robotics, Italian Institute of Technology, Genova, Italy

Antonio Di Lallo
Lab of Biomechatronics and Intelligent Robotics (BIRO), Department of Mechanical Engineering, The City University of New York, City College, NY, United States

Christian Di Natali
Department of Advanced Robotics, Italian Institute of Technology, Genova, Italy

Nafiseh Ebrahimi
Advanced Robotic Manipulators ARM Lab, Department of Mechanical Engineering, College of Engineering, University of Texas at San Antonio UTSA, San Antonio, TX, United States

Rand Hidayah
Robotics and Rehabilitation Laboratory, Mechanical Engineering Department, Columbia University, New York, NY, United States

Tzu-Hao Huang
Lab of Biomechatronics and Intelligent Robotics (BIRO), Department of Mechanical Engineering, The City University of New York, City College, NY, United States

Amir Jafari
Advanced Robotic Manipulators ARM Lab, Department of Mechanical Engineering, College of Engineering, University of Texas at San Antonio UTSA, San Antonio, TX, United States

Ravi Kant Jain
CSIR-Central Mechanical Engineering Research Institute (CMERI), Durgapur, India; Academy of Scientific and Innovative Research (AcSIR), Ghaziabad, India

Kwang J. Kim
Department of Mechanical Engineering, University of Nevada-Las Vegas, Las Vegas, Nevada, NV, United States

Samuel C.K. Lee
Biomechanics and Movement Science Program, University of Delaware, Newark, DE, United States; Department of Physical Therapy, University of Delaware, Newark, DE, United States; Shriners Hospitals for Children, Philadelphia, PA, United States

Tatiana Luna
Robotics and Rehabilitation Laboratory, Mechanical Engineering Department, Columbia University, New York, NY, United States

Justin Neubauer
Department of Mechanical Engineering, University of Nevada-Las Vegas, Las Vegas, Nevada, NV, United States

Jesús Ortiz
Department of Advanced Robotics, Italian Institute of Technology, Genova, Italy

Paul Schimpf
Department of Computer Science, Eastern Washington University (EWU), Cheney, WA, United States

Hao Su
Lab of Biomechatronics and Intelligent Robotics (BIRO), Department of Mechanical Engineering, The City University of New York, City College, NY, United States

Alexandrea Washington
Department of Mechanical Engineering, University of Nevada-Las Vegas, Las Vegas, Nevada, NV, United States

Shuangyue Yu
Lab of Biomechatronics and Intelligent Robotics (BIRO), Department of Mechanical Engineering, The City University of New York, City College, NY, United States

Introduction and acknowledgments

The robotic technology that is required for rehabilitation applications is fundamentally different that the one developed for industrial applications. In industrial applications, the structure of the robotic platforms has to be rigid, as any flexibility within the robot's body would lead to inaccuracy in position control and oscillation occurring in periodic motions which indeed would lead to errors in positioning the end effectors. This rigidity in the structure makes the entire platform bulky and heavy. Robotic arms with high rigidity and inertia that are being used in industrial applications have to move fast, which leads to very dangerous, sometimes deadly scenarios, in the case of physical interaction with humans.

Generally, rehabilitation focuses on aspects such as training intensity, repetitious practice, and task specificity, in dealing with soft and delicate bodies of patients. Therefore softness is the key in robotic rehabilitation applications. Soft robotics is an emerging field in robotic technology which is gaining interest among the researchers. The subject of this book is the current comprehensive research and development of soft robotic platforms for rehabilitation applications.

Chapter 1, High-Performance Soft Wearable Robots for Human Augmentation and Gait Rehabilitation, discusses high-performance soft wearable robots for human augmentation and gait rehabilitation. This chapter goes through the principles behind this new enabling technology and provides demonstrations of its beneficial application in several wearable robots.

Chapter 2, Development of Different Types of Ionic Polymer Metal Composite-Based Soft Actuators for Robotics and Biomimetic Applications, introduces the development of different types of ionic polymer metal composite (IPMC)-based soft actuators for robotics and biomimetic applications. This development method of fabrication of IPMC soft actuators shows significant advantages, such as the easy process of fabrication using nontoxic materials. Further, it is also proven that the development of IPMC soft actuators-based robotic systems provides better opportunities during compliant robotic assembly.

Chapter 3, Soft Actuators and Their Potential Applications in Rehabilitative Devices, talks about soft actuators and their potential applications in rehabilitative devices. There are different types of soft actuators that can be used, including Electroactive polymers (EAPs), ionic electroactive polymers (ionic EAPs), and hybrid actuators (HAs). These actuators can be used during a patient's daily activity and can even be worn on the body for consistent and convenient rehabilitation.

Chapter 4, Design Optimization of a Solenoid-Based Electromagnetic Soft Actuator With Permanent Magnet Core, introduces the design optimization of a solenoid-based electromagnetic soft actuator (ESA) with a permanent magnet core. Design optimization of the coil is discussed considering the geometrical parameters of the coil, including its length, inner and average diameters, number of turns, and packing density while the power consumption is bounded.

Chapter 5, Cable-Driven Systems for Robotic Rehabilitation, discusses cable-driven systems for robotic rehabilitation. Cable-driven systems are versatile, as they are inherently compliant to many types of human motion. Persons with impairments such as stroke, spinal cord injury, and cerebral palsy have varied motor control pathologies that cannot always be accommodated by rigid-link exoskeletons due to the imposition of restraints on the user's natural movement.

Chapter 6, XoSoft: Design of a Novel Soft Modular Exoskeleton, introduces XoSoft and presents design of a novel soft modular exoskeleton. XoSoft is a novel device that exploits new concepts of quasipassive actuation principles combined with smart technologies for sensors and actuators.

Chapter 7, TwAS: Treadmill with Adjustable Surface Stiffness, introduces Treadmill with Adjustable Surface Stiffness (TwAS). TwAS is integrated with the following systems: a weight cancelation hydraulically actuated harness system LiteGait, a motion capture system with eight Vicon cameras, and a ParvoMedics VO$_2$max metabolic cost measurement system. With this integrated system, a mobility impaired patient can walk at different speeds on different surface stiffness levels, while the walking kinematics can be tracked by the motion capture system and the energy expenditure can be measured by the VO$_2$max system.

Finally Chapter 8, An Artificial Skeletal Muscle for Use in Pediatric Rehabilitation Robotics, introduces an artificial skeletal muscle for use in pediatric rehabilitation robotics. Five soft actuator candidates that can be deployed in exoskeleton applications, are compared in this chapter, including coiled nylon fiber (CNF), ethanol-based phase-change (EPC), polymer actuators, plasticized poly vinyl chloride (PVC) gel, a stacked dielectric elastomer (DE) actuator, and a hydraulically amplified self-healing electrostatic (HASEL) actuator.

This book is perfect for graduate students, researchers, and professional engineers in robotics, control, mechanical, and electrical engineering who are interested in soft robotics, artificial intelligence, rehabilitation therapy, and medical and rehabilitation device design and manufacturing.

Special thanks go to program officers at the Disability and Rehabilitation Engineering (DARE) program at the National Science Foundation (NSF) and Prof. Hai-Chao Han, Department Chair of the Mechanical Engineering Department at the University of Texas at San Antonio (UTSA) for their support and help.

Editors
Amir Jafari and Nafiseh Ebrahimi
September 2020

CHAPTER 1

High-performance soft wearable robots for human augmentation and gait rehabilitation

Antonio Di Lallo[1], Shuangyue Yu[1], Tzu-Hao Huang[1], Thomas C. Bulea[2] and Hao Su[1]

[1]Lab of Biomechatronics and Intelligent Robotics (BIRO), Department of Mechanical Engineering, The City University of New York, City College, NY, United States
[2]Functional and Applied Biomechanics Section, Rehabilitation Medicine Department, Clinical Center, National Institutes of Health, Bethesda, MD, United States

1.1 Introduction

Recent years have seen the development of an increasing number of powered exoskeletal devices for a widening range of applications. Several designs have been proposed either to enhance strength and endurance capabilities in able-bodied subjects [1] or to assist impaired movements in disabled people [2–4].

From a design perspective, wearable robots can be generally classified as rigid or soft in terms of actuation and transmission. Although rigid exoskeletons allow provision of the strongest assistance, on the other hand it is recognized that excessive mass and high impedance represent two key drawbacks of such robots [5]. In the course of addressing this issue, several designs of soft exoskeletons using pneumatics have been proposed [6]. However, pneumatic actuation typically relies on tethered air compressors, making its application for portable systems still challenging. Therefore lately soft cable-driven textile exosuits have become the new trend in wearable robots research [7]. Some examples include exosuits for ankle [8] and hip [2] joint assistance during walking. Thanks to their conformal and compliant structure, these devices are unobtrusive and extremely lightweight, but the absence of rigid links makes their design significantly challenging. The main issues are related to the difficulty of a fixed positioning, especially corresponding to certain joints (e.g., knee), and to the unavoidable presence of shear forces that usually result in annoyance for the wearer.

Therefore trying to fill the gap between rigid exoskeletons and soft exosuits, hybrid solutions enable excellent compromises in terms of light weight, compliance, applied forces, and allowable range of motion. Remarkable performance is made possible thanks to the employment of a new actuation paradigm, purposely designed for highly dynamic interactive tasks. It constitutes a step forward toward the development of "physically intelligent" robots and will reveal opportunities for effective human−robot interaction.

The rest of the chapter will discuss how this new actuation paradigm allows dynamic interactive tasks when it is employed to power wearable robots. In Section 1.2 the involved actuation technologies are described, while Section 1.3 presents their application to three kinds of exoskeletons, for hip, knee, and back assistance. Finally, conclusions are discussed in Section 1.4.

1.2 Actuation technologies for physical human−robot interaction

Safe and dynamic interaction with humans is of paramount importance for collaborative robots. Recent exoskeletons have focused on advanced algorithms to improve control performance [2,9−13], while there is limited work in actuation hardware design.

Yet, observations from nature have highlighted that body properties of biological organisms have a key function in providing "physical intelligence" during dynamic interactive tasks. In the same manner, mechanical design of robots' hardware represents an opportunity to provide embedded intelligence, that is, a sort of intelligence intrinsic to the system without relying on complicated control algorithms [14].

When dealing with wearable robots, indeed actuators play the greatest role among the various mechanical components.

Besides actuator prototypes for soft material robots, state-of-the-art wearable robots include three primary actuation methods, namely, conventional geared actuation, series elastic actuation (SEA), and quasi-direct drive actuation (QDD, also known as proprioceptive actuation) [14−16]. Some significant examples are reported in Fig. 1.1.

Actuation paradigm	Conventional actuation	Series elastic actuation	Conventional actuation and textile wearable	Quasi direct drive actuation
	High gear ratio; Convectional motor; Load and wearable	High ratio gearbox; Load and wearable; Convectional motor; Spring, encoder	High gear ratio; Convectional motor; Load and wearable	High torque density motor; Load and wearable; Low ratio gearbox
Torque (Nm)	70 ✓	60 ✓	32 ~	48 ✓
Mass (kg)	23 ✗	7 ~	5 ~	3 ✓
Bandwidth (Hz)	5 ✗	5 ✗	20 ✓	44 ✓
Compliance (Nm)	30 ✗	9 ~	N/A ~	0.4 ✓
Torque density (Nm/kg)	3 ✗	8.6 ~	6.4 ~	16 ✓

Figure 1.1 Performance comparison of selected exemplifying embodiments of the main actuation methods employed in wearable robots [14]. They include conventional geared actuators, in both rigid [17] and textile exosuits [9,18], series elastic actuators [19,20], and quasi-direct drive actuators [15].

Conventional actuation uses high-speed and low-torque motors (typically brushless direct current motors, BLDC) coupled to high gear ratio transmission [3,17,21,22]. It can meet the typical requirements related to assistive torque, angular velocity, and control bandwidth, but suffers from high mechanical impedance, which makes the overall system quite resistant to free movements of the wearer. Although control algorithms might be able to partially compensate for the undesirable impedance, a complete suppression of the obtrusive effects due to the high inertia of the actuator remains challenging.

Series elastic actuators (including parallel elastic actuators and other variable stiffness elastic actuators) overcome the low-compliance limitation [19,20,23,24] using spring-type elastic elements. However, this solution is detrimental to the system simplicity, weight, and size, and leads to a sacrifice of the performance in control bandwidth, resulting in limited practical benefits for wearable robots.

QDD actuation instead involves a change of perspective. Being composed of high-torque motors and low gear ratio transmission, it exhibits intrinsic high torque density, high bandwidth, and high backdrivability, meeting all the multifaceted requirements of versatile wearable robots. Therefore these features identify QDD actuation as a promising technology for dynamic human–robot interaction, whose feasibility has already been proven in several preliminary tethered exoskeletons [7,25–27]. Other recent results are reported in Section 1.3 about wearable robots for hip, knee, and back assistance.

1.2.1 High torque density motors

QDD actuation is a new paradigm of robot actuation design that leverages high torque density motors with low ratio transmission mechanism [14,28]. It has been recently studied for legged robots [14] and exoskeletons [29].

The benefits of high torque density actuation include a simplified mechanical structure with reduced mass and volume, and high compliance, that is, high backdrivability. Thus it is an ideal candidate to satisfy static and dynamic requirements of wearable robots.

A crucial component for the design of high torque density actuation is the high torque density motor. In [30] a custom BLDC motor is designed with optimized mechanical structure, topology, and electromagnetic properties. It uses high-temperature resistive magnetic materials and adopts an outer rotor and a flat, concentrated winding structure to maximize the torque density [31,32]. As shown in Fig. 1.2A, in order to enhance the efficiency the motor uses fractional-slot type winding, which allows it to reduce the cogging torque and to minimize the copper loss. It has 21 pole pairs and 36 rotor

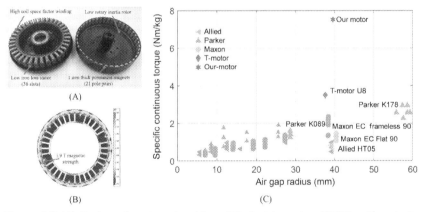

Figure 1.2 (A) Design of custom brushless DC electric motor (BLDC) with exterior rotor and concentrated winding. The fractional-slot design (36 slots, 21 pole pairs) allows it to reduce cogging torque and to minimize copper loss [33]. (B) Finite element analysis shows that by using sintered neodymium iron boron permanent magnets the magnetic strength of the stator can reach 1.9 T, under condition of no current in the winding. (C) Distribution of continuous torque density versus air gap radius for our custom motor compared to commercial ones. It is worth noting that exoskeletons typically need motors with air gap radius in the 35–40 mm range. Thanks to its ad hoc design, the custom motor developed at The City University of New York, marked with a star in the plot, exhibits continuous torque density (7.81 Nm/kg) 10.4 times higher than the Maxon brushless DC motor EC flat 90 (#323772, 0.75 Nm/kg), widely used in the exoskeleton industry.

slots, a number significantly higher than the 12 pole pairs of the commercial motor Maxon EC90 [23] or the 10 pole pairs and 18 rotor slots of another concurrent research prototype [33]. Moreover, unlike conventional BLDC motors that place windings around the rotor, here motor winding is attached to the stators and the rotor consists only of the cover and 1-mm-thick permanent magnet chips. In this way, the lightweight exterior rotor reduces rotary inertia and increases the torque to inertia ratio. Finally, a further detail not to be neglected involves the choice of the electromagnetic material: permanent magnets are made of sintered neodymium iron boron (NdFeB), which can reach 1.9 T magnetic field intensity, as shown by finite element analysis reported in Fig. 1.2B.

Overall, this design allows a significant reduction in the inertia and mechanical impedance of the motor while increasing its control bandwidth. It weighs 256 g and provides 2 Nm continuous torque. Fig. 1.2C shows the distribution of continuous torque density versus air gap radius for this motor compared to commercial ones [28]. In the 35−40 mm air gap radius domain, usually adopted for wearable robots, the continuous torque density of the described motor is 7.81 Nm/kg, remarkably higher than the values of the widely used T-motor U8 (3.5 Nm/kg) and Maxon EC Flat 90 (#323772, 0.75 Nm/kg).

1.2.2 Quasi-direct drive actuation

The high torque density motor constitutes an important step toward an effective actuation strategy for wearable robots. To keep the system compact and lightweight, it is worth combining the motor with all the related components in a fully integrated actuator, as shown in Fig. 1.3A [34,35].

The overall actuator weight is 777 g and it includes the high torque density motor, an 8:1 ratio planetary gear, a 14-bit high accuracy magnetic encoder, and a wide range input (10−60 V) motor driver and controller.

Detailed specifications of the resulting actuator can be found in Table 1.1, together with the corresponding values of a reference example from conventional actuators and series elastic actuators.

Low-level control on position, velocity, and current is implemented in the driver-control electronics, whereas real-time information for high-level control transfer is through the Controller Area Network (CAN bus) communication protocol.

When powered with a nominal voltage of 42 V, the actuator reaches a nominal speed of 188 RPM (19.7 rad/s). Moreover, thanks to the quasi-

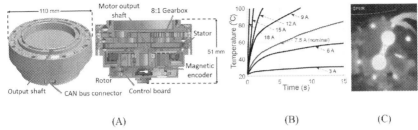

Figure 1.3 (A) Fully integrated quasi-direct drive actuator, including the high torque density motor, an 8:1 gearbox, a magnetic encoder, and control electronics. It is compact (Φ110 mm × 52 mm height), lightweight (777 g), and able to generate high torque (16 Nm nominal torque and 45 Nm peak torque). (B) Stator temperature over time under different current conditions. (C) Thermal image after 15 min of continuous 7.5 A current operation shows that the actuator surface reaches a temperature of 62.7°C.

Table 1.1 Performance comparison of the three main actuation paradigms.

Parameter	Unit	Conventional [21]	SEA [23]	QDD
Output rated torque	Nm	8	40	17.5
Actuator mass	kg	~0.50	1.80	0.77
Actuator rated torque density	Nm/kg	16	22.2	20.7
Control bandwidth	Hz	5.1	4.2	73.3
Backdrive torque	Nm	—	—	0.97

direct drive strategy using low gear ratio transmission, the actuator presents low output inertia (57.6 kg·cm^2), which means low resistance to natural human movements.

Regarding the output capability, it is worth noting that it is highly limited by the motor's winding temperature. To evaluate the actuator working current performance, it was operated continuously in stall mode under different output currents. The stator temperature was measured by an embedded temperature sensor and the surface temperature was measured by a portable FLIR thermal camera. The experiment was performed in a 22°C lab environment without external heat dissipation. The maximum operating time was set to 15 min and the maximum temperature of the stator to 100°C. In Fig. 1.3B the evolution over time of the stator temperature is plotted for different current conditions, while Fig. 1.3C shows the thermal camera image of the actuator surface after 15 min under 7.5 A nominal current, demonstrating that a highest temperature of 62.7°C is reached. This experiment validates that the actuator can produce a continuous output torque of 17.5 Nm under 7.5 A rated current.

1.3 Applications to wearable robots

Having discussed the properties of QDD actuation and its potential impact on applications involving dynamic human−robot interaction, this section will analyze in detail three different use cases: a portable exoskeleton for hip assistance during walking and squatting, and two tethered versions, one for knee support during squatting and the other for back assistance in stoop lifting.

1.3.1 Hip exoskeleton

During walking human hip joints have flexion/extension movements in the sagittal plane and abduction/adduction movements in the frontal plane. Therefore a hip exoskeleton needs to accommodate those two degrees of freedom. For level-ground walking, the range of motion of a human hip joint is 32.2 degrees flexion, 22.5 degrees extension, 7.9 degrees abduction, and adduction 6.4 degrees [36]. Here the robot is designed with a larger range of motion than the standard requirements to handle a heterogeneous population for a wide variety of activities beyond walking, such as squatting, sitting, and stair climbing. Moreover, it was observed that for a human of 75 kg walking at 1.25 m/s the peak torque and the speed of the hip joint are 97 Nm and 3.5 rad/s, respectively, but results in [37] and [38] show that a 12 Nm torque assistance is sufficient to produce a 15.5% reduction in metabolic cost for uphill walking.

1.3.1.1 Design

The mechanical system of the hip exoskeleton is symmetric about the sagittal plane and is mainly composed of a waist frame, two actuators, two torque sensors, and two thigh braces, as shown in Fig. 1.4. The waist frame has the main function of anchoring the actuators. It has a curvature conformal to the wearer's pelvis, enabling uniform force distribution on the human body. A wide waist belt is used to attach the waist frame to the user, aiming at maximizing the contact area so as to reduce the pressure on the human. The motor housings are connected to the waist frame by means of hinge joints that enable passive degrees of freedom in the frontal plane (e.g., abduction and adduction), whereas the actuators work in the sagittal plane to assist the flexion and extension of the hip joints. A

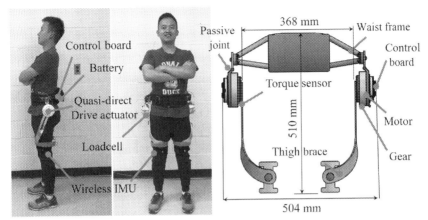

Figure 1.4 The hip exoskeleton is composed of a waist frame, two QDD actuators, two torque sensors, and two thigh braces. It provides assistance for flexion and extension of the hip joints in the sagittal plane, while passive joints allow free abduction and adduction movements.

customized compact torque sensor is assembled to the output flange of the actuator to measure the output torque. Finally, the thigh brace is fixed to the sensor and transmits the actuator torque to the wearer's thigh thanks to a fastening strap. In order to alleviate the wearer from painful shear forces, the thigh brace has a curved structure that enables it to provide assistive forces on the thigh perpendicularly to the frontal plane.

Regarding the electrical system, it supports high-level torque control, low-level motor control, sensor signal conditioning, data communication, and power management. The local motor controller is developed based on a motor driver and a DSP microcontroller. It allows measurement of the motor motion status and to realize control based on current, velocity and position. The high-level microcontroller runs on Arduino Due and performs torque control. It acquires real-time data on the lower-limb posture from the wireless inertial measurement unit (IMU) sensors and on the applied torques from the loadcells.

1.3.1.2 Modeling

The overall system can be broken down into four main subsystems, as shown in Fig. 1.5: the motor, the transmission mechanism, the wearable structure, and the human leg. Connections between these modules are represented as springs and dampers to model the force and motion transmission.

High-performance soft wearable robots for human augmentation and gait rehabilitation

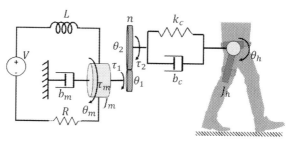

Figure 1.5 Human–exoskeleton coupled dynamic model. It consists of four subsystems: motor, transmission, wearable structure, and human leg.

The dynamics of the motor electrical system can be characterized by the winding resistance R and inductance L. Subject to input voltage V, the motor generates a torque τ_m proportional to the current i and to the torque constant k_t, whereas the back-electromagnetic force V_b is proportional to the motor velocity $\dot{\theta}_m$ and to the constant k_b. Therefore the governing equations of the motor electrical system are

$$V - V_b = L\frac{di}{dt} + Ri \tag{1.1}$$

$$\tau_m = k_t i \tag{1.2}$$

$$V_b = k_b \dot{\theta}_m \tag{1.3}$$

Meanwhile, from a mechanical perspective the motor is described by the equation

$$\tau_m = J_m \ddot{\theta}_m + b_m \dot{\theta}_m + \tau_1 \tag{1.4}$$

where J_m denotes the moment of inertia of the rotor around its rotation axis, b_m is the damping coefficient that takes into account internal viscous friction, θ_m denotes the motor angle, and τ_1 is the torque applied to the output shaft.

Then, considering a gearbox with gear ratio $n:1$, it has the effect of reducing the angular velocity and amplifying the torque according to the equations

$$\theta_1 = \theta_m; \quad \theta_2 = \frac{\theta_1}{n}; \quad \tau_2 = n\tau_1. \tag{1.5}$$

In Eq. (1.5), θ_1 and θ_2 denote the rotation angles of the input and output shafts of the gearbox, respectively, and τ_1 and τ_2 the corresponding applied torques.

The wearable structure of the exoskeleton can be designed with rigid linkages, springs, cable—pulley systems, and cable—textile systems. In all the cases, it can be modeled through global parameters, namely stiffness k_c and damping b_c. Therefore the resulting equation for the dynamics of the wearable structure is

$$\tau_2 = b_c(\dot{\theta}_2 - \dot{\theta}_h) + k_c(\theta_2 - \theta_h). \tag{1.6}$$

Finally, the human limb is governed by the equation

$$J_t\ddot{\theta}_h + b_c(\dot{\theta}_h - \dot{\theta}_2) + k_c(\theta_h - \theta_2) = \tau_l, \tag{1.7}$$

where J_t is the inertia of the limb with the orthosis, θ_h is the hip rotation angle, τ_l is the human torque generated by the muscles, and τ_a is the torque applied on the human thigh. τ_a can be an assistive or resistive torque and can be calculated as

$$\tau_a = \tau_2 = b_c\left(\frac{\dot{\theta}_m}{n} - \dot{\theta}_h\right) + k_c\left(\frac{\theta_m}{n} - \theta_h\right). \tag{1.8}$$

Assume as initial condition $\theta_h(0)$, $\dot{\theta}_h(0)$, and $V(0)$ equal to zero and neglect the inductance L due to its small value [39]. Let s be the Laplace variable, in the s-domain the assistive torque $\tau_a(s)$ is related to the hip rotation angle $\theta_h(s)$ and input voltage $V(s)$ as expressed in Eq. (1.9).

$$\tau_a(s) = G_1(s)V(s) + G_2(s)\theta_h(s) = n_1(s)\left[\frac{n_2(s)}{d(s)}V(s) + \frac{n_3(s)}{d(s)}\theta_h(s)\right], \tag{1.9}$$

where

$$n_1(s) = (b_c s + k_c)$$

$$n_2(s) = nk_t$$

$$n_3(s) = -n^2\left[J_m R s^2 + (R b_m + k_b k_t)s\right] \tag{1.10}$$

$$d(s) = J_e s^2 + b_e s + k_e$$

$$J_e = n^2 J_m R; \ b_e = n^2 R b_m + n^2 k_b k_t + R b_c; \ k_e = R k_c.$$

The natural frequency ω_n of the open-loop torque control for the second-order system is

$$\omega_n = \sqrt{\frac{k_c}{J_e}} = \sqrt{\frac{Rk_c}{n^2 J_m R}} = \sqrt{\frac{k_c}{n^2 J_m}}. \quad (1.11)$$

The effective moment of inertia J_e is equal to $n^2 J_m$, hence the natural frequency ω_n of the open-loop torque control is directly proportional to wearable structure stiffness k_c and inversely proportional to the square of the gear ratio n and the moment of inertia of the motor J_m.

The described model allows also to make some considerations about the backdrivability. To identify the property of the passive mechanism, $V(s)$ is set to zero and the output resistive torque τ_a induced by the human motion $\theta_h(s)$ can be derived from Eq. (1.12), while Eq. (1.13) provides the output link impedance.

$$\tau_a(s) = G_2(s)\theta_h(s) = \frac{-(b_c s + k_c)n^2[J_m R s^2 + (Rb_m + k_b k_t)s]}{n^2[J_m R s^2 + (Rb_m + k_b k_t)s] + Rb_c s + Rk_c}\theta_h(s) \quad (1.12)$$

$$Z_o(s) = \frac{\tau_a(s)}{s\theta_h(s)} \quad (1.13)$$

As the gear ratio is sufficiently small, the resistive torque can be neglected, because

$$\lim_{n \to 0} \tau_a(s) = 0 \quad (1.14)$$

As the gear ratio is large enough, the resistive torque is approximated by Eq. (1.15), where the wearable structure damping b_c and stiffness k_c are the dominating terms.

$$\lim_{n \to \infty} \tau_a(s) \approx -(b_c s + k_c)\theta_h(s) \quad (1.15)$$

When the gear ratio is equal to 1, the resistive torque is expressed by Eq. (1.16). It depends on the gear ratio n, the damping term b_c, the stiffness k_c, the motor inertia J_m, the motor damping b_m, the motor resistance R, the motor torque constant k_t, and the back EMF constant k_b.

$$\tau_a(s)|_{n=1} = -\frac{(b_c s + k_c)(J_m R s^2 + Rb_m s + k_b k_t s)}{J_m R s^2 + (Rb_m + k_b k_t + Rb_c)s + Rk_c}\theta_h(s) \quad (1.16)$$

Therefore from Eqs. (1.14), (1.15), and (1.16) it is clear that high backdrivability (i.e., low resistive torque and low output impedance) can be achieved with small gear ratio n, small damping constant b_c, and small stiffness k_c.

1.3.1.3 Control

The control system is based on a hierarchical architecture composed of a high-level control layer that robustly detects gait intention (Fig. 1.6 top), a middle-level control that generates the assistive torque profile (Fig. 1.6 bottom), and a low-level control layer that implements a current-based torque control.

An algorithm based on a data-driven method [40] with a neural network regressor is used to compensate for the uncertainties caused by changing gait speeds. It estimates the walking and squatting cycle percentage in real-time by the signals from two IMUs mounted on the anterior of both thighs (Fig. 1.4). These sensors provide motion information, including Euler angles, angular velocities, and accelerations at a frequency of 200 Hz. Motion information during the last 0.4 s sliding time window constitutes the input vector of the neural network for both offline training process and online control. The neural network used in this algorithm has one hidden layer with 30 neurons as well as a sigmoid activation function and deploys the Xavier initialization [41] for the network weights. The

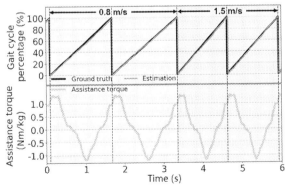

Figure 1.6 (Top) Estimated gait cycle percentage accurately matches the ground truth. The latter is calculated offline by insole signals, while the estimation is provided applying a regression method to information from IMUs. The robust gait recognition ($R^2 = 0.997$) is able to compensate for the disturbances due to walking speed changes. (Bottom) Assistive torque profile generated by the control algorithm.

algorithm could achieve an $R^2 = 0.997$ on a test set of walking and squatting data collected from three able-bodied subjects at several different speeds.

After obtaining the gait percentage, the middle-level controller calculates the assistive torque proportionally according to a predefined torque profile expressed as a look-up table. Basically, the desired assistive torque is obtained by searching the gait percentage in the look-up table and using interpolation to fill in the missing data. The predefined torque profile for walking is generated by the human biological model in [42], while the one for squatting is expressed as a simple sine wave.

Finally, the low-level torque control architecture is composed of an inner and an outer loop control. The inner loop implements motor current control in the local motor controller, while the outer loop performs torque control in Arduino Due using feedback signals from motors, loadcells, and IMU-based gait recognition.

1.3.1.4 Evaluation

Several experiments were conducted on the hip exoskeleton to characterize its mechanical versatility through backdrivability and bandwidth demonstrations.

For the bandwidth experiment chirp signals with different magnitudes were used as reference torque to obtain the Bode plot. Results are shown in Fig. 1.7A, where bandwidth values of 57.8, 59.3, and 62.4 Hz are obtained for 10, 15, and 20 Nm chirp magnitude, respectively. Thus the bandwidth is much higher than the requirement of human walking, but this property turns out to be useful for agile human activities, for example, running and balance control to unexpected external disturbance. Compared with the exoskeleton using SEA [19], characterized by 5 Hz bandwidth, a high control bandwidth robot is safer and more robust to uncertainties.

For the backdrivability experiment, instead, the backdrive torque was measured in unpowered mode. An angular displacement of 32.2 degrees was imposed to the hip joint at 1 Hz frequency while the actuator was turned off and the resistance torque was measured. The profiles of the rotation angle and the backdrive torque are reported in Fig. 1.7B. Results show that the hip exoskeleton presents a very low backdrive torque (maximum value is about 0.4 Nm), demonstrating higher compliance than other state-of-the-art exoskeletons [19,33].

14 Soft Robotics in Rehabilitation

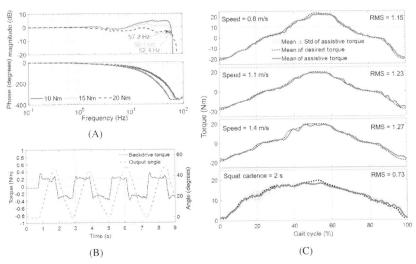

Figure 1.7 (A) Bode plot of the 10, 15, and 20 Nm torque control, demonstrating remarkably high control bandwidth. (B) Backdrive torque measured in unpowered mode for the imposed joint angular displacement. The maximum resistance torque is approximately 0.4 Nm, significantly lower than other state-of-the-art devices (e.g., 2 Nm in [19] and 1 Nm in [33]). (C) Torque tracking performance of assistance (peak torque is limited to ±20 Nm) during walking and squatting tests. Mean of actual assistive torque (solid line) is able to track the desired torque (dashed line) with high accuracy. RMSEs of torque tracking (0.8 m/s, 1.1 m/s, and 1.4 m/s walking, 2 s cadence squatting) are 1.15, 1.23, 1.27, and 0.73 Nm respectively (5.75%, 6.15%, 6.35%, and 3.65% of the peak torque).

As a last experiment, a control test was performed to investigate the torque tracking performance of the hip exoskeleton. It was tested during treadmill walking with varying speed from 0.8 to 1.4 m/s and during squatting with 2 s cadence. A total of 15 tests with the same torque profile were performed for each of the walking and squatting motions. The tracking performance of the hip assistance is shown in Fig. 1.7C. The average moot mean square error (RMSE) between the desired and actual torque trajectory in 60 tests is 1.09 Nm (5.4% of the maximum desired torque). This result indicates that the torque controller is able to track with high accuracy the desired assistance during walking and squatting.

1.3.2 Knee exoskeleton

Recently there is a growing interest in wearable robots for knee joint assistance as cumulative knee disorders account for 65% of lower extremity musculoskeletal disorders. Squatting and kneeling are two of the primary risk factors that contribute to knee disorders [43].

Knee joint assistance during squatting necessitates a broad range of motion (0–130 degrees flexion) and joint torque (up to 60 Nm) [44]. Moreover, for an effective synchronization with the wearer, the torque generated from the robot needs to be delivered at an angular velocity of no less than 2.4 rad/s.

1.3.2.1 Design

Most of the existing knee exoskeletons are designed for walking assistance [45,46] and they typically do not allow the squat motion due to the interference between the robot structure and human bodies (e.g., [47,48]). Since the focus of the work is to understand the feasibility of the approach for squat assistance, the exoskeleton consists of a wearable robot emulator, that is a tethered wearable structure with offboard actuation. Fig. 1.8A shows the overall system, including the wearable structure, a bidirectional Bowden cable transmission mechanism and the high torque density actuator implemented as a tethered platform. It is worth noting that though the current platform is configured as a tethered system, it can be easily converted to a portable system, as the overall mass of motor and gears is 0.55 kg.

The bidirectional Bowden cable mechanism (similar to [49] and [50]) uses a single motor to generate bidirectional actuation, that is, knee flexion and extension. In this regard, a key role is played by the knee joint mechanism (Fig. 1.8B), that constitutes the distal portion of the bidirectional cable-drive mechanism. It is designed to be lightweight and low-profile, namely to avoid interference with the human body during squat motion.

The assembly includes one flexion cable and one extension cable that pass around the distal pulley and terminate at the cable locking mechanism. One side of the knee mechanism is attached to the thigh brace while the shank plate is fixed to the calf brace. A load cell connects the thigh and the calf links and plays a key role in force transmission between the cable and the shank plates. In fact, when the cable is pulled, it actuates the pulley through the locking mechanism and drives the shank plate via the loadcell.

The exoskeleton is attached to the body via 3D printed carbon fiber braces designed to be conformal to the human leg. Thanks to these braces the torque at the knee joint is converted into pressure distributed along the length of the thigh and the shank. Therefore the size of the wearable arms

16 Soft Robotics in Rehabilitation

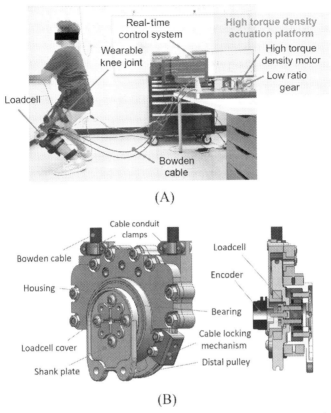

Figure 1.8 (A) The hip exoskeleton includes the wearable structure with the knee joint mechanism, a bidirectional Bowden cable transmission system, and a high torque density actuation platform. (B) Section view (left) and isometric view (right) of the cable-driven knee joint mechanism for bidirectional actuation, that is, knee flexion and extension.

plays a crucial role in the performance and user comfort. Three-dimensional infrared scans (Sense 2, MatterHackers Inc.) of the wearer's leg are taken and processed into a CAD model. This model is then 3D printed using fused deposition modeling with carbon fiber reinforcements. Foam padding is also added in the locations of leg contact to aid in comfort, while Velcro straps are used to anchor the exoskeleton arms to the leg.

1.3.2.2 Modeling

A human biomechanics model is derived to calculate the knee joint torque to assist both squat and stoop lifting activities in real time. Unlike methods that use simple and predefined profiles (e.g., sine waves) to approximate the human joint torque, this method is biologically meaningful and applicable to squat, stoop, and walking activities. In [51] an assistive algorithm for a squat assistance exoskeleton is proposed assuming that the back of the subject is straight, and the trunk angle is zero. It only uses knee joint angle to calculate the required torque and lacks the posture information of the hip and trunk. However, during lifting (squat and stoop) the back angle varies and significantly affects the knee joint torque.

Since squat and stoop involve significantly different biomechanics of the knee joint, this model is versatile in the sense that it can cover both scenarios for a wide variety of people. The knee joint torque $\hat{\tau}_k$ can be derived from Eq. (1.17)

$$\hat{\tau}_k = I(\theta)\ddot{\theta} + C(\theta,\dot{\theta})\dot{\theta} + G(\theta), \tag{1.17}$$

where θ is the joint angles, $I(\theta)$ is the inertia matrix, $C(\theta,\dot{\theta})$ denotes the centrifugal and Coriolis term, and $G(\theta)$ is the gravitational loading.

As typically lifting tasks are relatively slow, the knee joint torque is dominated by the gravitational loading. Thus, with reference to Fig. 1.9, estimated knee joint torque $\hat{\tau}_k$ can be computed using a quasi-static model, as expressed in Eq. (1.18).

$$\hat{\tau}_k = G(\theta) = -0.5[M_b g(L_b \sin\theta_b + L_t \sin\theta_t) + M_t g L_{tc} \sin\theta_t] \tag{1.18}$$

Here the knee extension is defined as the positive direction for the knee joint torque $\hat{\tau}_k$, while the clockwise direction is defined as the positive direction for the trunk angle θ_b, the thigh angle θ_t, and the shank

Figure 1.9 Quasi-static model used to derive the assistive knee joint torque during squat motion.

angle θ_s. M_b is the combined mass of the head, neck, thorax, abdomen, pelvis, arms, forearms, and hands, M_t is the mass of thigh, L_b is the length between the center of mass M_b and the hip pivot, L_t is the length of thigh between the hip and the knee pivots, L_{tc} is the length between the center of mass M_t and the knee pivot, and g is the gravitational constant. The parameters L_b, L_t, L_{tc}, M_b, M_t are calculated according to Eqs. (1.19–1.23) using data in Table 1.2 obtained from anthropometry research [52]. It is worth noting that the proposed model is customizable to different individuals because the assistive torque can be adjusted to the subject's weight and height by means of the weight ratio (ratio between the subject weight M_{sb} and the human model weight M_W) and the height ratio (ratio between the subject height L_{sb} and the human model height L_H).

$$M_b = \left(\frac{M_{sb}}{M_W}\right) \cdot \sum_{i=1}^{8} M_i \tag{1.19}$$

$$M_t = \left(\frac{M_{sb}}{M_W}\right) \cdot M_9 \tag{1.20}$$

Table 1.2 The human segment parameters.

#	Segment	M_i: Mass (kg) Total weight M_W: 81.4 kg	L_i: Length between center of mass to ground (m) Total height L_{Ht}: 1.784 m
1	Head	M_1: 4.2 kg	L_1: 1.679 m
2	Neck	M_2: 1.1 kg	L_2: 1.545 m
3	Thorax	M_3: 24.9 kg	L_3: 1.308 m
4	Abdomen	M_4: 2.4 kg	L_4: 1.099 m
5	Pelvis	M_5: 11.8 kg	L_5: 0.983 m
6	Arms	M_6: 4 kg	L_6: 1.285 m
7	Forearms	M_7: 2.8 kg	L_7: 1.027 m
8	Hands	M_8: 1 kg	L_8: 0.792 m
9	Thighs	M_9: 19.6 kg	L_9: 0.75 m
10	Calfs	M_{10}: 7.6 kg	L_{10}: 0.33 m
11	Feet	M_{11}: 2 kg	L_{11}: 0.028 m
12		Hip pivot to ground	L_{12}: 0.946 m
13		Knee pivot to ground	L_{13}: 0.505 m

$$L_b = \left(\frac{L_{sb}}{L_H}\right) \cdot \left\{ \left[\frac{\sum_{i=1}^{8}(M_i \cdot L_i)}{\sum_{i=1}^{8}(M_i)}\right] - L_{12} \right\} \quad (1.21)$$

$$L_t = \left(\frac{L_{sb}}{L_H}\right) \cdot (L_{12} - L_{13}) \quad (1.22)$$

$$L_{tc} = \left(\frac{L_{sb}}{L_H}\right) \cdot (L_9 - L_{13}) \quad (1.23)$$

Finally, given the estimated joint torque $\hat{\tau}_k$, the desired assistive torque τ_r to be provided by the exoskeleton is defined as

$$\tau_r = \alpha \cdot \hat{\tau}_k. \quad (1.24)$$

As long as the gain α is positive, the exoskeleton will assist the wearer. It can be used to reduce the loading and increase the endurance of workers. On the other hand, when the gain α is negative, the exoskeleton will resist the human. It can be useful to increase the muscle strength for healthy subjects in fitness activities or individuals with movement impairments in rehabilitation.

1.3.2.3 Control

The control system is based on a two-level configuration architecture, as shown in Fig. 1.10: a target computer is used for high-level assistive control, while local motor driver electronics perform low-level control. The control strategy follows the method adopted in [53], where the authors demonstrate accurate force tracking of a robot arm in contact with surfaces of unknown linear compliance. Here, the same strategy is adapted to control the interaction torque between exoskeleton and human. Unlike [11], where a predefined and fixed torque reference is used, the present control provides adaptive assistance to the wearer based on the biomechanics model for both squatting and stooping.

The high-level controller runs at 1 kHz and implements a torque loop proportional-integral-derivative (PID) scheme to track the reference assistive torque.

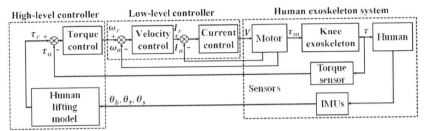

Figure 1.10 Block diagram of the assistance control algorithm. The high-level controller generates a reference torque profile based on the biomechanics model. τ, I, and ω denote the torque, the current, and the velocity, respectively. V and τ_m are the motor input voltage and output torque. θ_b, θ_t, and θ_s denote the trunk, thigh, and shank angles. Subscripts r and a refer to the reference and actual values, respectively.

The low-level controller instead implements a velocity loop PID scheme running at 20 kHz, and a current PID control running at 200 kHz. It measures the real-time motor status (i.e., current, velocity, and position) and communicates with the target computer through the CAN bus.

In addition, three IMU sensors, and one loadcell are connected to the computer through corresponding interface boards. The IMUs provide measurements of the trunk angle θ_b, thigh angle θ_t, and shank angle θ_s with a sampling rate of 400 Hz.

They are calibrated to zero degrees at the beginning of the experiment, while the subject is instructed to stand straight. Then, the knee angle θ_k and hip angle θ_h are calculated by Eqs. (1.25–1.26) and their positive directions represent an extension.

$$\theta_k = \theta_t - \theta_s \tag{1.25}$$

$$\theta_h = \theta_t - \theta_b \tag{1.26}$$

1.3.2.4 Evaluation

Several experiments were carried out to demonstrate the compliance of the exoskeleton, the control effectiveness, and the torque tracking performance.

The study was approved by the City University of New York Institutional Review Board, and all methods were carried out in accordance with the approved study protocol. Based on the experimental procedure, three healthy subjects performed five repetitions of squat motion. Rhythm was marked by a metronome and each cycle took 8 s, as shown in Fig. 1.11.

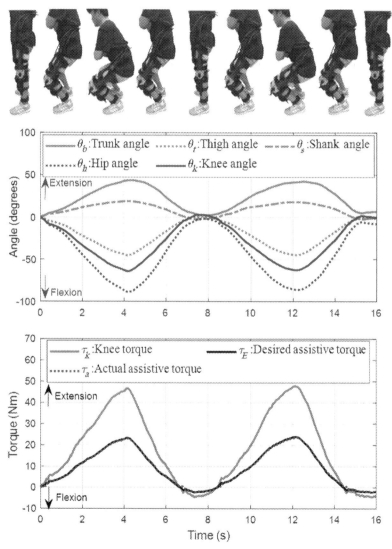

Figure 1.11 Squat assistance control strategy. Top graph plots the evolution over time of the trunk, hip, thigh, knee, and shank angles during two squatting cycles. Bottom graph reports the required knee joint torque, and both the desired and actual assistive torques in the case of 50% level of assistance.

For the compliance evaluation, the backdrive torque was measured during squatting in unpowered condition. Thanks to the high torque density motor, the low gear ratio transmission, and low-friction cable-drive mechanism, a very low mechanical impedance is obtained, as shown in Fig. 1.12A. The peak backdrive torque is registered at the onset of motor rotation and corresponds to the changes of direction. The average resistant torque is 0.92 Nm, while its maximum value is 2.58 Nm, much lower than other state-of-the-art knee exoskeletons (for instance, as an example case the corresponding peak resistance in [29] is 8 Nm). These results were confirmed also by the subjects, who reported extremely low resistance while wearing the device.

The same test was performed also with zero torque tracking control, whereby the resistant torque was measured while the actuator was turned on and the reference torque was steadily set to zero, regardless of human motion. This trial was implemented to compensate for the mechanical resistance, such as friction of the cable and gears. Accordingly, the mechanical impedance was further reduced compared to the unpowered condition, as shown in Fig. 1.12B. The average resistant torque is 0.34 Nm (4 times lower), while its maximum value is 0.64 Nm (2.7 times lower).

As a further evaluation, the torque tracking performance during squatting was analyzed for three levels of assistance, namely 10%, 30%, and 50% of the required knee joint torque calculated by Eq. 1.18, corresponding to $\alpha = 0.1$, $\alpha = 0.3$, and $\alpha = 0.5$ in Eq. 1.24, respectively. Therefore

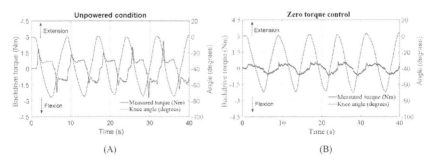

Figure 1.12 Characterization of the mechanical impedance during squatting in the case of unpowered condition (A) and zero torque tracking control (B). The orange line indicates the knee angle, while the blue one shows the measured resistant torque. The peak backdrive torque is registered at the onset of motor rotation and corresponds to the changes of direction. Average backdrive torque is 0.92 Nm in case (A) and 0.34 in case (B), while the peak values are 2.58 and 0.64 Nm, respectively.

the assistive control was used to augment human knee joints during squats by applying specific torque depending on the current trunk angle θ_b and thigh angle θ_t detected by the IMU sensors. The tracking performance is shown in Fig. 1.13.

The RMSE between the desired and actual torque trajectory is 0.23 Nm (2.8% of 7.6 Nm peak torque), 0.22 Nm (1.1% of 20 Nm peak torque), and 0.29 Nm (1.2% of 23.9 Nm peak torque) in 10%, 30%, and 50% knee assistance, respectively. These results demonstrate that the torque controller can deliver the desired torque profile with higher accuracy: the overall RSME of torque tracking is about 0.29 Nm (1.2% of the peak torque) while for instance in [54] it is 2.1 Nm (21% error of 10 Nm peak torque).

Finally, the effectiveness of the assistance provided by the exoskeleton was evaluated in terms of its capability to reduce muscle activity.

For this purpose, the knee extensors (rectus femoris, vastus lateralis, vastus medialis) and the knee flexors (biceps femoris and semitendinosus) EMG signals were observed in six different scenarios: without the

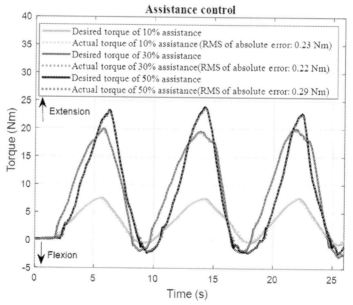

Figure 1.13 Tracking performance of 10%, 30%, and 50% knee torque assistance in three squatting cycles. The RMSE between the desired and actual torque trajectory is 0.23, 0.22, and 0.29 Nm, respectively. Overall RMSE of torque tracking is less than 0.29 Nm (1.21% of 24 Nm peak torque).

exoskeleton, power-off exoskeleton, zero torque control assistance, 10%, 30%, and 50% assistance.

Fig. 1.14A reports the data relative to the vastus lateralis of a single subject.

It shows that in the passive condition, due to the mechanical impedance of the wearable structure EMG amplitude of power-off condition is slightly higher than the one without exoskeleton. In the active condition the EMG amplitude of the zero torque control is pretty similar to the one without exoskeleton, while it is clearly reduced in 10%, 30%, and 50% assistance. Therefore these results reveal that the assistive control is effective in reducing the effort of the knee extensor muscle.

In an attempt to analyze the assistive effect in a more comprehensive way, Table 1.3 and Fig. 1.14B report the RMS amplitude of the EMG signals of the five observed muscles for each of the six conditions, whereby data are averaged over 15 squat cycles (five squat cycles per three subjects). EMG of

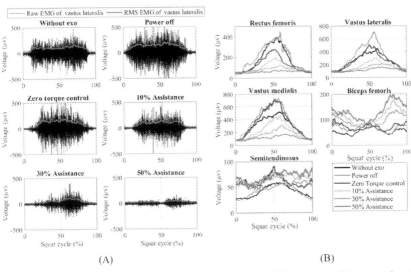

Figure 1.14 Muscle activities during squatting in six different conditions: without-exoskeleton, power-off, zero torque control, 10%, 30%, and 50% assistance. (A) The vastus lateralis EMG of a single subject is plotted. It reveals that the assistive control has the beneficial effect of reducing the effort of the vastus lateralis muscle. (B) EMG data of knee extensor (rectus femoris, vastus lateralis, and vastus medialis) and flexor (biceps femoris and semitendinosus) muscles during squatting in different conditions. Lines represent the average RMS EMG from 15 squat cycles (five squat cycles per three subjects). Results shows that due to exoskeleton assistance the activities of the knee extensor muscles were reduced, while those of flexor muscles were increased.

Table 1.3 Average RMS EMG in knee muscles.

Muscles		Sub.	WO	Off	0%	10%	30%	50%
Knee Extensors	Rectus femoris	S1	78	87	35	25	28	11
		S2	31	36	45	26	15	10
		S3	56	65	49	46	35	39
		Avg.	**55**	**62**	**43**	**32**	**26**	**20**
	Vastus lateralis	S1	123	149	108	50	47	26
		S2	44	51	61	39	20	11
		S3	89	124	81	79	72	78
		Avg.	**85**	**108**	**83**	**56**	**46**	**38**
	Veastus medialis	S1	87	124	111	50	46	14
		S2	98	117	135	78	38	12
		S3	112	129	101	98	88	89
		Avg.	**99**	**124**	**116**	**75**	**54**	**38**
	Knee extensors		80	98	80	55	42	30
Knee Flexors	Biceps femoris	S1	23	25	23	63	43	57
		S2	13	20	18	13	16	20
		S3	20	25	32	37	37	55
		Avg.	**19**	**23**	**24**	**38**	**32**	**44**
	Semitendinosus	S1	13	14	11	20	15	18
		S2	14	19	19	14	14	15
		S3	15	14	30	31	24	36
		Avg.	**14**	**16**	**20**	**22**	**18**	**23**
	Knee flexors		16	19	22	30	25	34

Unit (μV); WO (without exo); Off (power off); 0% (zero torque); 10% (10% assistance); 30% (30% assistance); 50% (50% assistance).

knee extensors (rectus femoris, vastus lateralis, vastus medialis) reach the highest amplitude in power-off condition, but they are not significantly different to the cases without exoskeleton and with zero torque control. In particular, the values of standard deviation between these two conditions are 0.0188, 0.0136, and 0.0246 for rectus femoris, vastus lateralis, and vastus medialis, respectively. Meanwhile, it is clear that the higher the torque delivered to the wearer, the lower is the muscle activity of the knee extensors. However, an increase of the muscle activities of knee flexors (biceps femoris and semitendinosus) is observed. As these are bi-articular muscles, which extend the hip and flex the knee, their increased activation is possibly required to stabilize the pelvis in response to the assistance. Another reason for this effect may be found in the lack of training by the novice users of the exoskeleton. Further investigation will be needed for a precise clarification.

In summary, experimental results indicate that the proposed exoskeleton is highly backdrivable with minute mechanical resistance and that moderate levels of assistance can effectively reduce muscle effort during squatting. In particular, it was observed that the proposed exoskeleton can reduce the knee extensors activity, but it is still not clear if the work is globally alleviated or simply transferred to adjacent muscle groups (e.g., hip extensors, hip flexors, ankle extensors, and ankle flexors), due to the complex mechanism of muscle group compensation. Metabolics measurements will be used in the future to perform a more in-depth analysis of the actual efficacy.

1.3.3 Back exoskeleton

Back injuries are the most prevalent work-related musculoskeletal disorders [55]. Wearable robots present an attractive solution to mitigate ergonomic risk factors and reduce musculoskeletal loading for workers who perform lifting. Over the last two decades, various studies have demonstrated that industrial exoskeletons can decrease total work, fatigue, and load while increasing productivity and work quality [1,56]. For instance, Toxiri et al. developed a powered back-support exoskeleton that reduced 30% muscular activity at the lumbar spine [4], while a passive back exoskeleton with a larger range of motion of the trunk was proposed in [57]. The key challenges of back-support exoskeletons lie in the unique anatomy of the human spine, composed of 23 intervertebral discs. Therefore this structure imposes stringent requirements that necessitate new solutions for an effective human–robot interaction.

To address the aforementioned challenge, a spine-inspired continuum soft exoskeleton has been developed with the aim of reducing spine loading during stoop lifting while not limiting the natural movements. In particular, the stoop lifting induces extension and flexion of the lumbar joints with 70 degrees in the sagittal plane. Moreover, natural range of motion allows lateral flexion of 20 degrees in the frontal plane and rotation of 90 degrees in the transverse plane. Biomechanics analysis reveals that 250 N of the exoskeleton force perpendicular to the back can decrease 30% of the lumbar compression force at the lumbosacral joint (L5/S1, between fifth lumbar and first sacral) while a 15 kg load is lifted.

1.3.3.1 Design
Due to the requirement of having a system conformal to the human back anatomy and unobtrusive for the natural movements, the proposed robot

leverages a hyperredundant continuum structure that is able to continuously bend [58,59], as shown in Fig. 1.15A.

Additionally, Fig. 1.15B shows the overall setup, which includes a wearable structure made of shoulder and waist braces, the high torque density actuator implemented as a tethered platform, the Bowden cable transmission, and the control system. The spinal structure is a cable-driven mechanism and has a modular architecture composed of 20 segments. Each segment comprises a disc that pivots on a ball and socket joint. Thus each pair of neighboring disks forms a three-DOF spherical joint. A cable is threaded through holes at the edges of the discs, so that when the actuator pulls the cable, the discs rotate about the ball joint, acting as levers and producing assistive torque on the human. The electric motor delivers 2 Nm nominal torque at 1500 rpm nominal speed, and it is coupled to a gearbox with 36:1 gear ratio. As a result, the actuation platform can output up to 1500 N pulling force at 0.22 m/s cable translating speed. A customized load cell placed at the bottom of the spinal structure allows measurement of the cable tension. Moreover, an elastic backbone made of coiled steel tubing ensures a tight coupling of the various segments and the integration of the overall mechanism.

1.3.3.2 Modeling

A kinematics analysis is carried out to optimize the geometrical design and to characterize the range of motion of the mechanism.

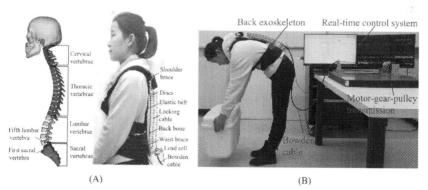

Figure 1.15 (A) The spine-inspired back exoskeleton leverages on a hyperredundant continuum mechanism. Thanks to its compliance, this wearable robot provides assistive force while being conformal to the anatomy of the human back and unobtrusive for natural motion. (B) A healthy subject wearing the back exoskeleton to perform stoop lifting of a 15 kg load. The spine continuum mechanism is powered by a tethered actuation platform via Bowden cable transmission.

The configuration of the back exoskeleton is determined by the accumulated rotations of all discs, as shown in Fig. 1.16A.

The pose of the $(i+1)$-th disc with respect to the i-th disc can be represented by the homogeneous transformation

$$T_{i+1} = \text{Rot}_x(\varphi_{i+1})\text{Rot}_y(\theta_{i+1})\text{Rot}_z(\psi_{i+1})\text{Tran}(l), \quad (1.27)$$

where $\varphi_{i+1}, \theta_{i+1}, \psi_{i+1}$ are the rotation angles of disc $i+1$ with respect to disc i in the sagittal, frontal and transverse planes, respectively, and $l = (0\ 0\ 1)^T$ is the distance vector between two neighboring discs. $\text{Rot}_x(\cdot), \text{Rot}_y(\cdot), \text{Rot}_z(\cdot)$ denote 4×4 homogeneous transformation matrices representing rotations around x, y, and, z axes, respectively, while $\text{Tran}(\cdot)$ is a 4×4 homogeneous translation matrix.

By putting all of this together, the global pose transformation of the mechanism from the base to the distal disc n can be calculated by

$$T_{tot} = T_1 T_2 \cdots T_n. \quad (1.28)$$

From Eq. (1.28) we see that the overall range of motion is the accumulation of the ranges of motion of individual discs. The range of motion of one disc with respect to the adjacent disc depends on the geometric parameters of the disc and the spherical joint in between, as depicted in. When the disc rotates from the initial configuration represented in Fig. 1.16B to the extreme configuration due to a mechanical contact

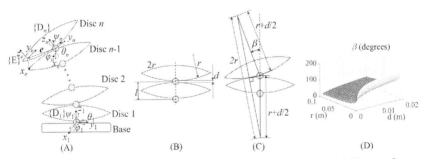

Figure 1.16 Kinematics analysis of the spine continuum structure. (A) The configuration of the exoskeleton is determined by the accumulated rotations of all the discs. (B) Initial configuration of two adjacent discs. (C) Extreme configuration of two adjacent discs due to a mechanical contact constraint. (D) Variation of β (maximal rotation angle between two neighboring discs) with respect to the geometric parameters r (radius of the disc) and d (distance between two neighboring discs). β affects the range of motion of the exoskeleton.

constraint, as depicted in Fig. 1.16C, the maximal rotation angle β can be calculated by

$$\beta = \pi - 2\arcsin\left(\frac{r}{(r + (d/2))}\right). \tag{1.29}$$

where r denotes the radius of the disc, d is the distance between two neighboring discs, and l is the distance between the centers of two neighboring spherical joints.

Accordingly, the range of motion, related to β, can be designed by adjusting the parameters r and d. Fig. 1.16D shows the effect of these two parameters on β.

For the present design it was chosen that $r = 0.07 m$ and $d = 0.00216 m$ to obtain $\beta = 20$, that allows to satisfy all the motion requirements (i.e., forward flexion of 70 degrees in the sagittal plane, lateral flexion of 20 degrees in the frontal plane, and rotation of 90 degrees in the transverse plane) with a number of discs greater than 6.

Given the kinematics characterization of the robot mechanism, it is crucial to examine a biomechanics model of human−robot interaction to facilitate the development of assistive control of the soft exoskeleton.

The kinetic purpose of the back exoskeleton is to reduce the compression and shear forces between discs, which are the main causes of low back pain. Therefore a basic analytical model of the forces acting on the human spine is derived to predict the effectiveness of the exoskeleton assistance on reducing the forces in the human spine and muscles.

For the sake of simplicity, the lumbar spine is modeled as a localized joint at the lumbar−sacral interface (L5/S1). Then, consider the condition when the human is in the flexed forward position during stoop lifting, as illustrated in Fig. 1.17. The static equilibrium analysis provides the relationship between the exoskeleton assistance and the forces in the human spine:

$$F_e D_e = -F_{exo} D_{exo} + m_{load} g D_{load} + m_{body} g D_{body} \tag{1.30}$$

$$F_p = F_e + m_{body} g \cos\theta + m_{load} g \cos\theta \tag{1.31}$$

$$F_s = -F_{exo} + m_{body} g \sin\theta + m_{load} g \sin\theta \tag{1.32}$$

F_p, F_s denote the compressive and shear forces of intervertebral discs, F_{exo} is the force applied by the back exoskeleton, and F_e denotes the force

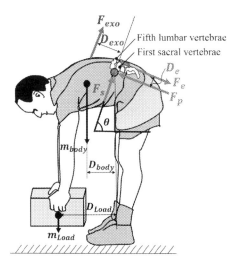

Figure 1.17 Biomechanics model of human–robot interaction during stoop lifting. If the exoskeleton applies a force perpendicular to the human back, it has the effect of accordingly reducing the spine compression force, the intervertebral shear force, and the lumbar muscle force.

of the erector spinae muscle. m_{body} and m_{load} are the masses of the human upper body and of the load, respectively. D_{exo}, D_e, D_{load}, and D_{body} are the moment arms of the exoskeleton, erector spinae muscle, load, and upper body, respectively.

According to Eqs. (1.30)–(1.32), it can be observed that if the exoskeleton force F_{exo} increases, the erector muscle force F_e, the spine compressive force F_p, and the intervertebral shear force F_s decrease simultaneously, because the weights of the human and of the load are partly balanced by the assistive force of the exoskeleton.

1.3.3.3 Control

The control architecture, as shown in Fig. 1.18A, consists of two main layers: a high-level controller and a low-level controller.

In the high-level controller a virtual impedance model, represented in Fig. 1.18B, is used to generate the reference assistive force according to Eq. (1.33).

$$F_r = \frac{T_r}{r_1} = \frac{1}{r_1}\left[J_d(\ddot{\theta}_a - \ddot{\theta}_r) + B_d(\dot{\theta}_a - \dot{\theta}_r) + K_d(\theta_a - \theta_r)\right], \qquad (1.33)$$

where θ_r, $\dot{\theta}_r$, and $\ddot{\theta}_r$ denote the desired trunk angle, velocity, and acceleration, generated from a predefined desired trajectory, while θ_a, $\dot{\theta}_a$, and $\ddot{\theta}_a$

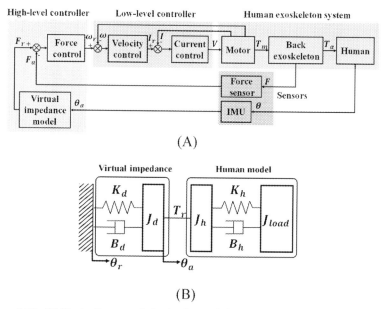

Figure 1.18 (A) Block diagram of the back exoskeleton control architecture for stoop assistance. It consists of two main controllers: the high-level controller receives sensor measurements about the cable force and the human trunk motion and generates the reference assistive force through the virtual impedance model, while the low-level controller implements motor velocity and current control. (B) Virtual impedance model. The assistive torque is generated by Eq. (1.33) from the desired reference position trajectory and the actual position trajectory with desired stiffness K_d, damping B_d, and inertia J_d. Using the virtual impedance model, the exoskeleton generated an assistive torque reference T_r.

are the actual values measured by an IMU sensor mounted on the trunk. In this case the desired trajectory is set to zero, so that virtual spring and damper are fixed to the ground.

High-level control is implemented in Matlab/Simulink Real Time and operates at 1000 Hz frequency. A PID force control is used to ensure that the cable measured force tracks the reference force F_r.

In the low-level controller, a DSP microcontroller (TMS320F28335, Texas Instruments, United States) is used for motor current and velocity control. It uses CAN bus communication to receive the desired velocity command V_r and to send data about the actuator state. Both velocity and current controllers implement a PID algorithm to track the reference signals.

Regarding the sensing system, a data acquisition (I/O) card (ADC, PCIe-6259, National Instrument, Inc., United States) is used to acquire cable force measurements from the loadcell mounted on the back exoskeleton, while an IMU mounted on the subject trunk transmits the trunk motion data (angle, angular velocity, and angular acceleration) via serial port (RS-232) to the target computer.

1.3.3.4 Evaluation

The exoskeleton provides assistance in stoop lifting without limiting natural motion. As shown in Fig. 1.19, the wearer is free to perform forward flexion, lateral flexion, and rotation.

Fig. 1.15B shows the setup used for the experimental evaluation of the back exoskeleton. Besides the wearable structure, it includes the Bowden cable transmission, the tethered actuation platform, and the real-time control system. Currently, a tethered actuation system is employed to perform a proof of concept trial, aimed at demonstrating the feasibility of the proposed spine design and control algorithm, thus minimizing the impact of the mass of the system. However, it is worth highlighting that the combined mass of motor and gearbox is just 0.55 kg, hence a portable version is indeed a practicable advancement already under development.

Three subjects performed 10 repetitions of 15 kg stoop lifting. Each stoop cycle took 8 s: 4 s for bending forward from stand-up posture to trunk flexion and 4 s for extending back from trunk flexion to stand-up posture. The study was approved by the City University of New York

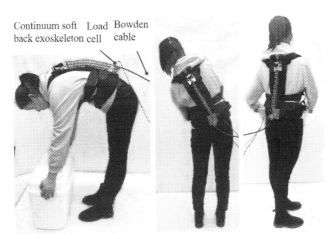

Figure 1.19 The continuum soft exoskeleton assists human stoop lifting while imposing no constraints on human forward flexion (left), lateral flexion (middle), and rotation (right).

Institutional Review Board, and all methods were carried out in accordance with the approved study protocol.

The first test regarded the steerability evaluation of the continuum exoskeleton, that is the relation between the cable displacement and the bending angle of the back exoskeleton, defined as the angle between the end faces of the base and the top disc.

Results shown in Fig. 1.20 indicate that a cable displacement of 5.23 cm produces a bending angle of 100 degrees, which is beyond the required range of motion of 70 degrees.

Then, the tracking performance of the assistive force control is evaluated. The desired assistance is calculated according to the virtual impedance model described by

$$F_r = 20\dot{\theta}_a + 200\sin\theta_a, \qquad (1.34)$$

where the sine function in the stiffness term has the function of compensating the involved components (which are related to $\sin\theta_a$) of the human and load gravity terms. Fig. 1.21 illustrates the variation of the force and the trunk angle during the stoop task, as observed from a total of 30 stoop cycles executed by three different subjects.

The RMSE of force tracking is 6.63 N (3.3% of the 200 N peak force). Therefore regardless of motion variability (represented by the standard deviation of trunk angles), the implemented controller was able to successfully track the desired force with high accuracy.

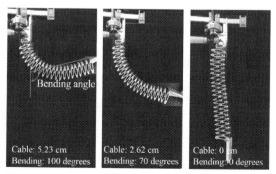

Figure 1.20 Steerability sequence of the continuum exoskeleton. The bending angle is defined as the angle between the end faces of the base and the top disc. A cable displacement of 5.23 cm is sufficient to produce a bending angle of 100 degrees.

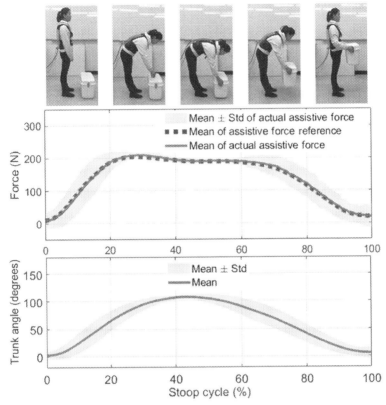

Figure 1.21 Assistive force tracking performance and trunk angle measurement during stoop lifting. Tests were executed by three healthy subjects and each subject performed 10 stoop cycles, for a total of 30 stoop repetitions. The mean actual assistive force (solid line) is able to accurately track the mean reference assistive force (dashed line). The shaded area identifies the variation within ± 1 standard deviation. RMSE of force tracking is 6.63 N (3.3% of the 200 N peak force).

1.4 Discussion

This chapter presented the advanced QDD actuation paradigm for high-performance wearable robots. Based on an ad hoc customized motor and low ratio gear transmission it ensures great versatility thanks to high torque density (20.7 Nm/kg), high backdrivability (0.4 Nm backdrive torque in unpowered mode), and high bandwidth (62.4 Hz). These properties are well suitable for applications involving human—robot interaction. Taking advantage of these characteristics three exoskeletons have been designed to provide assistance to the hip, the knee, and the back. Their feasibility and effectiveness were experimentally tested on healthy subjects. All of them exhibited low mechanical impedance and high accuracy in assistive force tracking, being able to overcome the performance of analogous state-of-the-art devices. In particular, the bilateral hip exoskeleton achieved 0.4 Nm backdrive torque, 62.4 Hz bandwidth, and RMSE in force tracking equal to 5.4% of 20 Nm peak torque. The bilateral knee exoskeleton presented backdrive torque equal to 1.5 Nm in unpowered mode and 0.5 Nm with zero torque tracking control, while RMSE of torque tracking was 1.2% of 24 Nm peak torque. Finally, for the spine exosuit RMSE of force tracking was about 3.3% of the 200 N peak force. In conclusion, experimental results verify that the presented actuation paradigm offers promising features to push the limits of wearable robots' performance. QDD actuation constitutes an enabling technology that could pave the way to the development of more lightweight, more compliant, safer, and stronger exoskeletons for either rehabilitation or augmentation purposes.

Acknowledgments

This work was supported by the National Science Foundation (NSF) grant CAREER CMMI 1944655, and National Institute on Disability, Independent Living, and Rehabilitation Research (NIDILRR) grant 90DPGE0011. T. C. Bulea was supported by the Intramural Research Program of the National Institutes of Health (NIH) Clinical Center. Any opinions, findings, and conclusions or recommendations expressed in this material are those of the author(s) and do not necessarily reflect the views of the funding organizations.

References

[1] M.P. De Looze, T. Bosch, F. Krause, K.S. Stadler, L.W. O'Sullivan, Exoskeletons for industrial application and their potential effects on physical work load, Ergonomics 59 (5) (2016) 671–681.
[2] Y. Ding, M. Kim, S. Kuindersma, C.J. Walsh, Human-in-the-loop optimization of hip assistance with a soft exosuit during walking, Sci. Robot. 3 (15) (2018) eaar5438.
[3] Z.F. Lerner, D.L. Damiano, T.C. Bulea, A lower-extremity exoskeleton improves knee extension in children with crouch gait from cerebral palsy, Sci. Transl. Med. 9 (404) (2017) eaam9145.
[4] S. Toxiri, A.S. Koopman, M. Lazzaroni, J. Ortiz, V. Power, M.P. de Looze, et al., Rationale, implementation and evaluation of assistive strategies for an active back-support exoskeleton, Front. Robot. AI 5 (2018) 53.
[5] J.C. Mcleod, S.J.M. Ward, A.L. Hicks, Evaluation of the keeogo dermoskeleton, Disabil. Rehabil.: Assist. Technol. 14 (5) (2019) 503–512.

[6] S. Sridar, P.H. Nguyen, M. Zhu, Q.P. Lam, P. Polygerinos, Development of a soft-inflatable exosuit for knee rehabilitation, 2017 IEEE/RSJ International Conference on Intelligent Robots and Systems (IROS), IEEE, 2017, pp. 3722–3727.
[7] J.S. Sulzer, R.A. Roiz, M.A. Peshkin, J.L. Patton, A highly backdrivable, lightweight knee actuator for investigating gait in stroke, IEEE Trans. Robot. 25 (3) (2009) 539–548.
[8] B.T. Quinlivan, S. Lee, P. Malcolm, D.M. Rossi, M. Grimmer, C. Siviy, et al., Assistance magnitude versus metabolic cost reductions for a tethered multiarticular soft exosuit, Sci. Robot. 2 (2) (2017) 1–10.
[9] J. Kim, G. Lee, R. Heimgartner, D.A. Revi, N. Karavas, D. Nathanson, et al., Reducing the metabolic rate of walking and running with a versatile, portable exosuit, Science 365 (6454) (2019) 668–672.
[10] E. Martini, S. Crea, A. Parri, L. Bastiani, U. Faraguna, Z. McKinney, et al., Gait training using a robotic hip exoskeleton improves metabolic gait efficiency in the elderly, Sci. Rep. 9 (1) (2019) 1–12.
[11] B. Chen, L. Grazi, F. Lanotte, N. Vitiello, S. Crea, A real-time lift detection strategy for a hip exoskeleton, Front. Neurorobot. 12 (2018) 17.
[12] F.L. Haufe, A.M. Kober, K. Schmidt, A. Sancho-Puchades, J.E. Duarte, P. Wolf, et al., User-driven walking assistance: first experimental results using the myosuit, 2019 IEEE 16th International Conference on Rehabilitation Robotics (ICORR), IEEE, 2019, pp. 944–949.
[13] A. Martnez, B. Lawson, M. Goldfarb, Á controller for guiding leg movement during overground walking with a lower limb exoskeleton, IEEE Trans. Robot. 34 (1) (2017) 183–193.
[14] P.M. Wensing, A. Wang, S. Seok, D. Otten, J. Lang, S. Kim, Proprioceptive actuator design in the mit cheetah: impact mitigation and high-bandwidth physical interaction for dynamic legged robots, IEEE Trans. Robot. 33 (3) (2017) 509–522.
[15] S. Yu, T.-H. Huang, D. Wang, B. Lynn, D. Sayd, V. Silivanov, et al., Design and control of a high-torque and highly backdrivable hybrid soft exoskeleton for knee injury prevention during squatting, IEEE Robot. Autom. Lett. 4 (4) (2019) 4579–4586.
[16] S. Seok, A. Wang, M. Yee Michael Chuah, D. Jin Hyun, J. Lee, D.M. Otten, et al., Design principles for energy-efficient legged locomotion and implementation on the mit cheetah robot, IEEE/ASME Trans. Mechatron. 20 (3) (2014) 1117–1129.
[17] J.L. Contreras-Vidal, N.A. Bhagat, J. Brantley, J.G. Cruz-Garza, Y. He, Q. Manley, et al., Powered exoskeletons for bipedal locomotion after spinal cord injury, J. Neural Eng. 13 (3) (2016) 031001.
[18] J. Bae, C. Siviy, M. Rouleau, N. Menard, K. O'Donnell, I. Geliana, et al., A light-weight and efficient portable soft exosuit for paretic ankle assistance in walking after stroke, 2018 IEEE International Conference on Robotics and Automation (ICRA) (2018) 2820–2827.
[19] I. Kang, H. Hsu, A. Young, The effect of hip assistance levels on human energetic cost using robotic hip exoskeletons, IEEE Robot. Autom. Lett. 4 (2) (2019) 430–437.
[20] X. Li, Y. Pan, G. Chen, H. Yu, Multi-modal control scheme for rehabilitation robotic exoskeletons, Int. J. Robot. Res. 36 (5-7) (2017) 759–777.
[21] Y. Lee, S.-g Roh, M. Lee, B. Choi, J. Lee, J. Kim, et al., A flexible exoskeleton for hip assistance, 2017 IEEE/RSJ International Conference on Intelligent Robots and Systems (IROS), IEEE, 2017, pp. 1058–1063.
[22] T. Lenzi, M.C. Carrozza, S.K. Agrawal, Powered hip exoskeletons can reduce the user's hip and ankle muscle activations during walking, IEEE Trans. Neural Syst. Rehabil. Eng. 21 (6) (2013) 938–948.
[23] T. Zhang, H. Huang, A lower-back robotic exoskeleton: industrial handling augmentation used to provide spinal support, IEEE Robot. Autom. Mag. 25 (2) (2018) 95–106.

[24] N. Paine, S. Oh, L. Sentis, Design and control considerations for high-performance series elastic actuators, IEEE/ASME Trans. Mechatron. 19 (3) (2013) 1080–1091.
[25] J.M. Caputo, S.H. Collins, Prosthetic ankle push-off work reduces metabolic rate but not collision work in non-amputee walking, Sci. Rep. 4 (1) (2014) 1–9.
[26] G. Lee, Y. Ding, I.G. Bujanda, N. Karavas, Y.M. Zhou, C.J. Walsh, Improved assistive profile tracking of soft exosuits for walking and jogging with off-board actuation, 2017 IEEE/RSJ International Conference on Intelligent Robots and Systems (IROS), IEEE, 2017, pp. 1699–1706.
[27] X. Yang, T.-H. Huang, H. Hu, S. Yu, S. Zhang, X. Zhou, et al., Spine-inspired continuum soft exoskeleton for stoop lifting assistance, IEEE Robot. Autom. Lett. 4 (4) (2019) 4547–4554.
[28] Y. Ding, H.-W. Park, Design and experimental implementation of a quasi-direct-drive leg for optimized jumping, 2017 IEEE/RSJ International Conference on Intelligent Robots and Systems (IROS), IEEE, 2017, pp. 300–305.
[29] G. Lv, R.D. Gregg, Underactuated potential energy shaping with contact constraints: application to a powered knee-ankle orthosis, IEEE Trans. Control. Syst. Technol. 26 (1) (2017) 181–193.
[30] J. Wang, X. Li, T.-H. Huang, S. Yu, Y. Li, T. Chen, et al., Comfort-centered design of a lightweight and backdrivable knee exoskeleton, IEEE Robot. Autom. Lett. 3 (4) (2018) 4265–4272.
[31] T. Reichert, T. Nussbaumer, J.W. Kolar, Torque scaling laws for interior and exterior rotor permanent magnet machines. IEEE International Magnetics Conference 2009 (INTERMAG 2009), A A 3 (2009) 1.
[32] J.J. Lee, W.H. Kim, J.S. Yu, S.Y. Yun, S.M. Kim, J.J. Lee, et al., Comparison between concentrated and distributed winding in ipmsm for traction application, 2010 International Conference on Electrical Machines and Systems, IEEE, 2010, pp. 1172–1174.
[33] H. Zhu, C. Nesler, N. Divekar, M.T. Ahmad, R.D. Gregg, Design and validation of a partial-assist knee orthosis with compact, backdrivable actuation, 2019 IEEE 16th International Conference on Rehabilitation Robotics (ICORR), IEEE, 2019, pp. 917–924.
[34] J. Wolff, C. Parker, J. Borisoff, W.B. Mortenson, J. Mattie, A survey of stakeholder perspectives on exoskeleton technology, J. Neuroeng. Rehabil. 11 (1) (2014) 169.
[35] D.J. Gonzalez, H.H. Asada, Hybrid open-loop closed-loop control of coupled human–robot balance during assisted stance transition with extra robotic legs, IEEE Robot. Autom. Lett. 4 (2) (2019) 1676–1683.
[36] W.E. Woodson, B. Tillman, P. Tillman, Human Factors Design Handbook: Information and Guidelines for the Design of Systems, Facilities, Equipment, and Products for Human Use, McGraw-Hill, 1992.
[37] K. Seo, J. Lee, Y.J. Park, Autonomous hip exoskeleton saves metabolic cost of walking uphill, 2017 International Conference on Rehabilitation Robotics (ICORR), IEEE, 2017, pp. 246–251.
[38] J. Lee, K. Seo, B. Lim, J. Jang, K. Kim, H. Choi, Effects of assistance timing on metabolic cost, assistance power, and gait parameters for a hip-type exoskeleton, 2017 International Conference on Rehabilitation Robotics (ICORR), IEEE, 2017, pp. 498–504.
[39] J. Malzahn, N. Kashiri, W. Roozing, N. Tsagarakis, D. Caldwell, What is the torque bandwidth of this actuator? 2017 IEEE/RSJ International Conference on Intelligent Robots and Systems (IROS), IEEE, 2017, pp. 4762–4768.
[40] J. Yang, T.-H. Huang, S. Yu, X. Yang, H. Su, A.M. Spungen, et al., Machine learning based adaptive gait phase estimation using inertial measurement sensors, 2019 Design of Medical Devices Conference, American Society of Mechanical Engineers Digital Collection, 2019.

[41] X. Glorot, Y. Bengio, Understanding the difficulty of training deep feedforward neural networks, Proceedings of the Thirteenth International Conference on Artificial Intelligence and Statistics, Chia Laguna Resort, 2010, pp. 249−256.
[42] A.S. McIntosh, K.T. Beatty, L.N. Dwan, D.R. Vickers, Gait dynamics on an inclined walkway, Journal of Biomechanics 39 (13) (2006) 2491−2502.
[43] C.R. Reid, P. McCauley Bush, N.H. Cummings, D.L. McMullin, S.K. Durrani, A review of occupational knee disorders, J. Occup. Rehabil. 20 (4) (2010) 489−501.
[44] V. Bartenbach, M. Gort, R. Riener, Concept and design of a modular lower limb exoskeleton, 2016 6th IEEE International Conference on Biomedical Robotics and Biomechatronics (BioRob), IEEE, 2016, pp. 649−654.
[45] T. Bacek, M. Moltedo, C. Rodriguez-Guerrero, J. Geeroms, B. Vanderborght, D. Lefeber, Design and evaluation of a torque-controllable knee joint actuator with adjustable series compliance and parallel elasticity, Mechanism Mach. Theory 130 (2018) 71−85.
[46] N. Karavas, A. Ajoudani, N. Tsagarakis, J. Saglia, A. Bicchi, D. Caldwell, Tele-impedance based stiffness and motion augmentation for a knee exoskeleton device, 2013 IEEE International Conference on Robotics and Automation, IEEE, 2013, pp. 2194−2200.
[47] J.E. Pratt, B.T. Krupp, C.J. Morse, S.H. Collins, The roboknee: an exoskeleton for enhancing strength and endurance during walking, IEEE International Conference on Robotics and Automation, 2004. Proceedings. ICRA'04. 2004, vol. 3, IEEE, 2004, pp. 2430−2435.
[48] K.A. Witte, A.M. Fatschel, S.H. Collins, Design of a lightweight, tethered, torque-controlled knee exoskeleton, 2017 International Conference on Rehabilitation Robotics (ICORR), IEEE, 2017, pp. 1646−1653.
[49] T.L. Nguyen, S.J. Allen, S.J. Phee, Direct torque control for cable conduit mechanisms for the robotic foot for footwear testing, Mechatronics 51 (2018) 137−149.
[50] K. Kong, J. Bae, M. Tomizuka, Torque mode control of a cable-driven actuating system by sensor fusion, J. Dyn. Syst. Meas. Control. 135 (3) (2013).
[51] A. Gams, T. Petric, T. Debevec, J. Babic, Effects of robotic knee exoskeleton on human energy expenditure, IEEE Trans. Biomed. Eng. 60 (6) (2013) 1636−1644.
[52] G. Harry, Armstrong. Anthropometry and mass distribution for human analogues, Military Male Aviat. 1 (1988).
[53] J. Roy, L.L. Whitcomb, Adaptive force control of position/velocity controlled robots: theory and experiment, IEEE Trans. Robot. Autom. 18 (2) (2002) 121−137.
[54] I. Kang, H. Hsu, A.J. Young, Design and validation of a torque controllable hip exoskeleton for walking assistance, ASME 2018 Dynamic Systems and Control Conference, American Society of Mechanical Engineers Digital Collection, 2018.
[55] Liberty Mutual Insurance, The Most Serious Workplace Injuries Cost Us Companies 59.9 Billion per Year, According to 2017 Liberty Mutual Workplace Safety Index, 2017.
[56] B.R. da Costa, E.R. Vieira, Risk factors for work-related musculoskeletal disorders: a systematic review of recent longitudinal studies, Am. J. Ind. Med. 53 (3) (2010) 285−323.
[57] M.B. Näf, A.S. Koopman, S. Baltrusch, C. Rodriguez-Guerrero, B. Vanderborght, D. Lefeber, Passive back support exoskeleton improves range of motion using flexible beams, Front. Robot. AI 5 (2018) 72.
[58] R.J. Webster III, B.A. Jones, Design and kinematic modeling of constant curvature continuum robots: a review, Int. J. Robot. Res. 29 (13) (2010) 1661−1683.
[59] S. Yu, T. Huang, X. Yang, J. Chunhai, J. Yang, Y. Chen, J. Yi, H. Su, Quasi-Direct Drive Actuation for a Lightweight Hip Exoskeleton with High Backdrivability and High Bandwidth, IEEE/ASME Transactions on Mechatronics (2020).

CHAPTER 2

Development of different types of ionic polymer metal composite-based soft actuators for robotics and biomimetic applications

Ravi Kant Jain[1,2]
[1]CSIR-Central Mechanical Engineering Research Institute (CMERI), Durgapur, India
[2]Academy of Scientific and Innovative Research (AcSIR), Ghaziabad, India

2.1 Introduction

In the recent past, soft robotics has been showing much interest toward the development of electroactive polymers (EAPs) soft actuators. An ionic polymer metal composite (IPMC) is one class of EAP (ionic polymer type) soft actuator. This is basically softly driven by a low voltage and can work quickly to provide bending responses [1–4]. The bending response can be utilized in different robotic and biomimetic applications such as robotic assembly, miniature parts assembly, biomimetic mechanics [5–9], aerospace, and medical applications [10,11], because they have light weight, easy processing, flexibility, high sensitivity, resilience, and biocompatible properties. They can also be used as artificial muscle-like actuators for various human affinity and biomedical applications [12,13]. To cater for such needs, IPMCs are therefore showing promise for the development/manufacturing of IPMC soft actuators for different applications. Principally, IPMCs consist of a thin ionomer composite polymer membrane with metal (Pt or Au) electrodes deposited on both faces. The protons on the anionic groups covalently bonded to the backbone of composite polymer membrane are typically exchanged for metal cations and the element is soaked with a solvent, usually water [14,15]. Consequently, the polymeric composite material-based actuators suffer associated problems such as short cycle lifetimes and lower response times [16–18]. Therefore it is required to use electrolytes with a polymeric membrane so that the performance of soft actuators can be enhanced. In this chapter, two different methods for the development of IPMC are proposed: (1) Kraton/graphene oxide/Ag/polyaniline (Kraton/GO/Ag/Pani) polymer composite-based soft actuators; and (2) sulfonated polyvinyl alcohol (SPVA) and 1-ethyl-3-methylimidazolium tetrachloro aluminate (IL) with deposited Pt composite-based soft actuators. These provide several advantages in their characteristics such as high proton conductivity (PC), high ion exchange, water uptake (WU), excellent film forming capacity, and large bending response. During development of the ionic polymer composite, the nonvolatile materials are used and characterized by their high ionic conductivities. These will provide benefits for the large actuation performance of IPMC actuators in air. Thus various combined chemical and electromechanical properties are studied in this chapter, which may improve the bending rate, displacement, and repeatability in the developed IPMC membrane. This book chapter is focused on the following points:

1. The fabrication of novel soft actuators based on Kraton/GO/Ag/Pani composite membrane and the SPVA and 1-ethyl-3-methylimidazolium tetrachloro aluminate (IL) with deposited Pt composite.
2. Chemical, mechanical, and electromechanical characterizations of different IPMC-based soft actuators.
3. Development IPMC soft actuators-based robotic system for compliant robotic assembly.

2.2 Literature survey on IPMC as actuators and sensors and its applications

In the last two decades, several researchers have carried out research work on the development, fabrications, modeling, control, and development of IPMC applications in robotic and biomimetic applications. Shahinpoor et al. [19] introduced IPMC as smart actuators and sensors along with the fundamental and mathematical modeling. Further, the use of IPMC as biomimetic sensors and actuators has shown large bending behavior and flapping displacement. Bar-Cohen [20] have focused development of artificial muscles using EAP, and their capabilities, challenges, and potential have been addressed. Further, Bar-Cohen [21] has shown the potential for developing IPMC actuators with performance characteristics that can be used as artificial muscles with the emergence of effective EAPs. Jung et al. [22] have developed a wireless tadpole robot that has a simple geometry, that is driven by low voltage and the undulatory fin-motion using IPMC actuators. The behavior of TadRob has been tested and analyzed under various frequencies (1–8 Hz) to find the correlation between actuator frequency and velocity of the robot. Yamakita et al. [23] have developed an IPMC-based artificial muscle linear actuator that gives a bending response to electric stimuli and this bending response is applied to robotic applications, especially to a biped walking robot. Kim et al. [24] have developed a wireless undulatory tadpole robot using an IPMC actuator and a biomimetic undulatory IPMC motion of the fin tail has been implemented. Tondu [25] has focused on the use of different types of IPMC as artificial muscles for humanoid robots and its characterization have been carried out for robotic applications. Kamamichi et al. [26] have focused on the doping effects of IPMC actuators for robotic applications where the doping effects have been demonstrated by dynamic motion or walking of a small-sized biped robot. Kim et al. [27] have summarized the EAP as artificial muscles actuators and sensors for robotic applications. Chen et al. [28] have focused on a control-oriented and physics-based model of IPMC actuators design. The model is represented as an infinite-dimensional transfer function relating the bending displacement to the applied voltage. Further, Chan et al. [29] have developed a physics-based model for biomimetic robotic fish propelled by an IPMC actuator. The biological fin structure is developed using a passive plastic fin and the IPMC actuator is attached to the tail's end. Koo et al. [30] have focused on the control of IPMC actuators using a self-sensing method which can be used for precise motion control. This IPMC actuator can maintain position during the handling of the robotic system. Najem [31] have attempted to design and develop a bioinspired robotic jellyfish using IPMC actuators. IPMCs are used as actuators in order to contract the bell and propel the jellyfish robot. Kim et al. [32] have focused on nonlinear learning control of IPMC where a controller is designed based on a reinforcement learning algorithm that can address probabilistic uncertainties and nonlinearity. Mutlu et al. [33] have attempted electromechanical modeling and parameter identification of IPMC actuators where the trilayer EAP as a soft robotic actuator model is developed by considering the significant number of rigid links connected with compliant revolute joints. Chen [34] has focused on a mathematical model of the adaptive control for IPMC actuators that is based on a continuous-time approach; a second-order dynamical system derived for a hysteresis compensation. Sun et al. [35] have combined the static and dynamic models of the

IPMC and derived the transfer function for the IPMC mechanical response to an electrical signal. This driving signal with a smooth slope and a low frequency is beneficial for the power efficiency. Wang et al. [36] have focused on the basic principle, fabrication process, and typical applications of IPMC and bucky gel actuator (BGA)-based actuators. Some real applications of IPMC and BGA are discussed. Peterson et al. [37] have explored the energy exchange between coherent fluid structures and IPMCs during impulsive loading of compliant IPMC strips. Fluid analysis has been carried out for flowing sensing and energy harvesting. Kaneto [38] has presented the use of EAP-based soft actuators using conductive polymers in robotic applications. Yan [39] has worked on the development, synthesis, and manufacturing of a novel ionic gel/metal nanocomposite (IGMN). Under this process, the supersonically accelerated gas-phase metal cluster beams are directed onto a polymeric substrate in order to generate thin conductive layers (few tenths to few hundreds of nanometers thick) anchored to the polymer. Carrico et al. [40] have provided a review of smart polymeric and gel actuator materials where an automated and freeform fabrication process, like 3D printing, is explored for creating the custom shaped monolithic devices. In particular, the advantages and limitations, manufacturing and fabrication techniques, and methods for actuator control are discussed. Further, Carrico et al. [41] have developed a 3D fused filament using an additive manufacturing (AM) technique in which electroactive polymer filament material is used to build soft active 3D structures layer by layer. Hong et al. [42] have studied the conjugated polymers where poly(3,4-ethylenedioxythiophene)-poly(styrenesulfonate) (PEDOT:PSS) are helpful to transfer the locally varying ion permeability of the ionic electroactive polymer actuators and manipulate ion motion to understand intrinsic angular deformation. Chen [43] presented a review on robotic fish enabled by the IPMC artificial muscles fabrication process for 3D actuating membranes, which has potential for a bioinspired robotic manta ray propelled by two IPMC pectoral fins. Pasquale et al. [44] have investigated the mechanical, thermal, and electromechanical behavior of IPMCs and IP2Cs at different solvents in order to study the actuation behavior. Shen [45] has focused on a multiple shape memory ionic polymer—metal composite (MSMIPMC) actuator with a multiple-shape memory effect which is able to perform complex motion by two external inputs, electrical and thermal. Theoretical and experimental investigations on the MSM-IPMC actuator are discussed. Khan et al. [46] have focused on sulfonated graphene oxide and sulfonated poly(1,4-phenylene ether-ether-sulfone) blended with polyvinylidene fluoride to create a IPMC actuator with enhanced performance for robotic and biomimetic applications. Ansaf et al. [47] have experimentally investigated the effects of humidity and actuation time on the electromechanical properties of IPMC to improve the practical applications for using the IPMCs as soft biomimetic sensors and actuators. Liao [48] have focused on a new class of artificial muscles based on the controllability of surface tension by electrocapillarity and surface oxidation with small voltages (0−1 V). Park et al. [49] have focused on the cost-effective P/(G−Ag) electrodes with high electrical conductivity that displayed a smooth surface resulting from the PEDOT:PSS coating, which prevents oxidation of the surface upon exposure to air, and shows strong bonding between the ionic polymer and the electrode surface. Nguyen et al. [50] have developed a trilayer conducting polymer bending microactuators using sequential stacking, solid polymer electrolyte, and micropatterning, where electrochemical and mechanical properties of the microactuators are studied. Annabestani et al. [51] have focused on a review of the IPMC as soft actuators and their applications in microfluidic micropumps, microvalves, and micromixers. Zhang et al. [52] have discussed the important characteristics and considerations in the selection, design, and implementation of various prominent and unique robotic artificial muscles for biomimetic robots, and provide perspectives on next-generation muscle-powered robots. Zhang et al. [53] have designed an ionic liquid gel (IGL) soft actuator and the material preparation principle was explained. Further the soft robot mechanism and the deformation mechanism of the ILG based on nonlinear finite-element theory are discussed.

Further, Zhang et al. [54] have focused on the motion simulation of ILG soft actuator which is based on the central pattern generator (CPG) control theory. The CPG-based bioinspired method can be used to control the soft robot drivers. Ando et al. [55] have attempted to develop low-order nonlinear finite-impulse response (NFIR) soft sensors for ionic electroactive actuators based on deep learning. The simulation results have been found using a soft sensor based on a fifth-order NFIR model. Kongahage et al. [56] have reviewed the use of smart material IPMC-based recent advances for smart textiles, where the incorporation of actuating materials provides a striking approach as a small change in the anisotropy properties of the materials. This enhances the significant performance due to the densely interconnected structures. Nguyen et al. [57] have designed an ionic exchangeable polymer electrolyte which shows actuation performance with a high displacement of 8.22 mm at a low voltage of 0.5 V and a fast rise time of 5 s. Sharma et al. [58] have presented the fabrication, characterization, and application of IPMC for robotic applications. Chen et al. [59] have proposed a new design of robotic fish which is propelled by a hybrid tail using IPMC. A state-space dynamic model is developed by capturing the two-dimensional (2D) motion dynamics of the robotic fish. Tabatabaie et al. [60] have presented the design for an artificial IPMC-based soft robot which has similar motions to an elephant's trunk. The bending deformation and the axial deformations of slit cylinders using IPMC are also discussed. Further, Tabatabaie et al. [61] have developed a new kind of IPMC actuators, energy harvesters, and sensors in slit cylindrical/tubular configurations. The slit IPMC cylindrical/tubular elements are made up of bending IPMC slit elements distributed symmetrically around a cylindrical mantle. MohdIsa et al. [62] have reviewed the IPMC sensing phenomenon, and the implementation and characteristics of different IPMC sensing methods where the sensing methods are divided into active sensing, passive sensing, and self-sensing actuation (SSA). The active sensing methods measure one of IPMC-generated voltage, charge, or current; passive methods measure variations in IPMC impedances, or use it in a capacitive sensor element circuit; and SSA methods implement simultaneous sensing and actuation on the same IPMC sample. Further, MohdIsa et al. [63] have studied and compared the frequency responses and noise dynamics of different IPMC active sensing signals, that is, voltage, charge, and current. By conducting experiments, the characteristics are identified by mechanically exciting IPMC samples, and simultaneously measuring the respective signals and material deformations.

2.3 Development of IPMC base soft actuator by different approaches

In this section, two different types of approach for the development of IPMC actuators are discussed:
1. Kraton/GO/Ag/Pani composite-based IPMC soft actuator
2. SPVA/IL/Pt IPMC composite-based IPMC soft actuator

The details for the Kraton/GO/Ag/Pani composite-based IPMC soft actuator are described below;

2.3.1 Kraton/GO/Ag/Pani composite-based IPMC soft actuator
2.3.1.1 Materials used
A nonperfluorinated polymer Kraton (pentablock copolymer poly((t-butyl-styrene)-b-(ethylene-r-propylene)-b-(styrene-r-styrene sulfonate)-b-(ethylene-r-propylene)-b-(t-butyl-styrene) (tBS-EP-SS-EP-tBS)); MD9200; Nexar Polymer, United States), silver nanopowder with a particle size range of 20–40 nm, tetrahydrofuran (THF) and aniline ($C_6H_5NH_2$; Thermo Fisher Scientific Pvt., Ltd., India), and ammonium peroxodisulfate [$(NH_4)_2S_2O_8$; extrapure Merck Specialties Pvt., Ltd., India] were used as received.

2.3.1.2 Fabrication of Kraton/GO/Ag/Pani composite-based IPMC soft actuator
The membrane fabrication was carried out by preparing 5 wt.% Kraton polymer solutions in THF at room temperature (25°C ± 3°C) with constant stirring up to 4 h. After complete dissolution, 0.3 mg silver nanopowder was added with constant stirring for 3 h at room temperature (25°C ± 3°C). Then, to the prepared solution, 2 mL of GO suspension was added with constant stirring for 2 h at 60°C followed by ultrasonication up to 1 h. The obtained solution was cast in a Petri dish covered with aluminum foil with small pores for slow evaporation of the solvent at room temperature. The membrane was taken out from the Petri dish and the surfaces were roughened using abrasive paper to increase the metal–polymer interface area, followed by washing with DMW and acetone to remove all impurities remaining from the manufacturing process.

The in situ oxidative polymerization of aniline on the Kraton/GO/Ag polymer membrane actuator was carried out by placing the dry membrane in (10% v/v in 1 M HCl) aniline solution in a conical flask using ammonium peroxodisulfate (0.1 M in 1 M HCl) as an oxidizing agent solution. Oxidant solution was then added slowly in the above solution to polymerize the aniline at the Kraton/GO/Ag membrane surface with constant stirring at a temperature below 10°C. As the oxidant was added, the color of the reaction mixture changed from purple to dark green within 15 min of stirring. After stirring for 1 h, the conical flask was covered with aluminum foil and kept in a refrigerator for 24 h for digestion. Finally, the Kraton/GO/Ag membrane was washed with DMW and methanol to remove the excess acid and dried at 45°C. The layer by layer coating of Pani on the surfaces of Kraton/GO/Ag ionomer membrane was carried out by repeating the above in situ oxidative polymerization process of aniline three times. After that, Kraton/GO/Ag/Pani-based Na^+-IPMC was

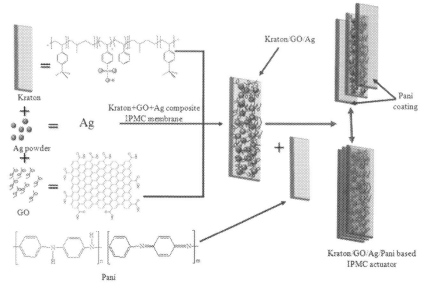

Figure 2.1 Chemical composition of Kraton/GO/Ag/Pani IPMC soft actuator.

obtained by placing the proposed H^+-IPMC in 0.2 M NaOH aqueous solution at room temperature for 6 h. Now, the Na^+-IPMC was washed with DMW followed by drying at 45°C for further characterization. Fig. 2.1 represents the systematic fabrication process of Kraton/GO/Ag/Pani membrane actuator.

2.3.1.3 Dehydration process

Dehydration process of the Kraton/GO/Ag/Pani IPMC membrane was analyzed with respect to time at room temperature. To test the dehydration process the sample was kept in distilled water at ambient temperature for 8 h to absorb water completely and measure the weight of the fully hydrated membrane. To determine the water loss from the IPMC membrane the weight was recorded every 2 min. The dehydration% of the membrane actuator was calculated using Eq. (2.1).

$$Dehydration\% = \frac{W_1 - W_2}{W_1} \times 100 \qquad (2.1)$$

where W_1 and W_2 are the weight of wet membrane and weight of membrane after dehydration at different interval of time.

2.3.1.4 Proton conductivity

PC of the IPMC sample (10 mm width, 30 mm length) was determined at room temperature by an impedance analyzer (FRA32M.X), connected with Autolab 302 N modular potentiostat/galvanostat, over a frequency of 50, 75, and 100 kHz under an AC perturbation of 50 mV/s. The PC of the membrane was calculated using Eq. (2.2):

$$\sigma = \frac{l}{R \times A} \quad (2.2)$$

where σ is PC in (S/cm), l is the thickness of the membrane in (cm), A is a cross-sectional area of membrane (cm^2), and R is the resistance (Ω).

2.3.1.5 Characterizations of Kraton/GO/Ag/Pani composite

The morphology, chemical and electromechanical properties, and chemical composition of the Kraton/GO/Ag/Pani composite-based IPMC were studied by a variety of techniques. The WU and ion exchange capacity (IEC) [64] of the proposed membrane were measured. The Fourier-transform infrared (FTIR) spectroscopy was carried out using a spectrometer (PerkinElmer Spectrometer, United States). Scanning electron microscopy (SEM) was carried out using SEM Jeol, JSM-6510LV, Japan. UV–visible absorption spectroscopy was employed to confirm the formation of Kraton/GO/Ag/Pani composite using Agilent Technologies Cart 60 UV–Vis, Malaysia. To determine the electrical property of the IPMC actuator, cyclic voltammetry (CV), linear sweep voltammetry (LSV) at triangle voltage input of ± 3 V with a step of 50–150 mV/s and impedance spectroscopy over a frequency range of 25, 50, and 75 kHz were performed with an Autolab 302 N modular potentiostat/galvanostat. For electromechanical characterization the maximum tip displacement of the proposed actuator was determined under sinusoidal voltage (± 4 V). The successive steps of bending response were also analyzed. For load characterization the IPMC membrane was clamped in a cantilever configuration and the maximum load carrying capability was determined. To find the repeatability of this actuator, several trials were conducted and the normal distribution was calculated.

2.3.2 SPVA/IL/Pt IPMC composite-based IPMC soft actuator
2.3.2.1 Materials used

4-Sulfophthalic acid (50 wt.% solution in water, technical grade, M.W. 246.19 g/mol) and 1-ethyl-3-methylimidazolium tetrachloro aluminate

(M.W. 279.96 g/mol) (Sigma- Aldrich Chemie Pvt., Ltd.), poly(vinyl alcohol) (PVA; M.W. ~115,000 g/mol, degree of polymerization 1700–1800, viscosity 25–32 cps, hydrolysis (mole%) 98–99) (Loba Chemie Pvt., Ltd., India), sodium nitrate ($NaNO_3$) (M.W. 84.99 g/mol), nitric acid (HNO_3), sodium hydroxide palette (M.W. 40 g/mol) and ammonium hydroxide (NH_4OH; 25%; Merck Specialties Pvt., Ltd., India), tetra-amineplatinum(II) chloride monohydrate [$Pt(NH_3)4Cl_2 \cdot H2O$ (crystalline) (M.W. 352.13 g/mol); Alfa Aesar], and sodium borohydride ($NaBH_4$) extra pure, (M.W. 37.83 g/mol; Sisco Research Laboratories Pvt., Ltd., India) were used as received without further purification.

2.3.2.2 Preparation of the reagent solutions

Aqueous solutions of $NaNO_3$ (1 M), HNO_3 (1 M), NaOH (0.1 M), Pt $(NH_3)_4Cl_2.H_2O$ (0.04 M), in double distilled water (DMW) were prepared. A solution of reducing agent was prepared by dissolving $NaBH_4$ in DMW with 5% (w/v) composition and 5% (v/v) NH_4OH aqueous solution was prepared from 25% NH_4OH solution.

2.3.2.3 Preparation of SPVA membrane

The preparations of membranes were started by dissolving a known amount of PVA in DMW with 10% (w/v) composition (i.e., the final volume of solution was 100 mL) under constant stirring for 6 h at 80°C. After complete dissolution, the solution was filtered using an ordinary cotton cloth. The benefits of using cotton cloth for filtration are (1) easy to filter without taking too much time and (2) the final concentration of the polymer solution remains the same. To the filtrate, 4 mL of 4-sulfophthalic acid (50 wt.% solution in water) and 200 mg 1-ethyl-3-methylimidazolium tetrachloro aluminate (IL) were added for sulfonation followed by continuous stirring overnight at 80°C. The homogeneous solution of the SPVA polymer was then cast into Petri dishes (100 × 15 mm (S line)) covered with aluminum foil with small pores for the slow evaporation of the solvent at 60°C in a hot air oven. After complete evaporation of the solvent, the membranes were taken out. Finally, the polymer membranes were cross-linked by placing them in a hot air oven at 120°C for up to 1 h. Then, the SPVA/IL polymer membranes were ready for electroplating of Pt metal.

2.3.2.4 Fabrication of SPVA/IL/Pt IPMC

A novel IPMC was fabricated by coating the Pt metal on both the surfaces of the SPVA/IL membrane using the electroless plating method. Electroless plating is the process of metalizing the polymer membrane surfaces by reducing the Pt metal ions into Pt metal particles using a reducing agent such as $NaBH_4$.

2.3.2.5 Characterizations of SPVA/IL/Pt IPMC

To check the physical and chemical properties of developed IPMC membrane actuator water-holding(WH) capacity, IEC, PC, water loss, and degree of sulfonation (%) (DS) were determined. A scanning electron microscope (SEM; JSM-6380; JEOL Ltd., Japan) was used to analyze the surface morphology and X-ray diffraction (XRD) of SPVA/IL/Pt IPMC membrane was recorded using Miniflex TM benchtop XRD framework (Rigaku Corporation, Tokyo, Japan) working at 40 kV and a current of 30 mA with Cu Kα radiation, wavelength $\lambda = 1.54$ A°, 1 scan/s, and scanning angle 20–80 degrees. FTIR spectrum of SPVA/IL/Pt ionic actuator was recorded between 500 and 4000 cm^{-1} using a spectrometer (PerkinElmer Spectrometer, United States). The electrical properties were carried out using CV at ± 3 V and LSV at 0–3 V with the help of an Autolab 302 N modular potentiostat/galvanostat at the scan rate 20–140 mV/s. For this, a three-electrode framework including SPVA/IL/Pt as working, an Ag/AgCl reference, and platinum wire as counter electrodes were utilized to obtain CV and LSV curves. For electromechanical characterizations, the successive steps of the bending deflection were measured using a laser displacement sensor (Model: OADM 20S4460/S14F; Baumer Electric, Germany) at 0–4 V DC through computer controlled digital analogue card (DAC) and microcontroller. RS-485 to RS-232 communication protocol was used in order to attain the proper communication between sensor and input with Docklight V1.8 software. To check the performance and repeatability of SPVA/IL/Pt IPMC membranes, multiple experiments were conducted to investigate the deflection hysteresis behavior. The deflection error of the fabricated IPMC was reduced by using a proportional integral derivative (PID) control system, where the controller bandwidth was set by tuning the frequency. To analyze the load characterization of the SPVA/IL/Pt IPMC a digital weighing/load cell (Model: Citizen CX-220, Make: India) was used. For this, the IPMC membrane was clamped in a cantilever configuration and the maximum load carrying capability was determined. Several trials of developed IPMC were conducted and the normal distribution was presented to find the repeatability and the stress behavior with respect to the applied potential. After electromechanical characterizations, a flexible link manipulator using SPVA/IL/Pt-based IPMC membrane was developed.

2.4 Results and discussions

In this section, the characterization of different types of IPMC actuators is discussed.

2.4.1 Characterization of Kraton/GO/Ag/Pani composite-based IPMC soft actuator

2.4.1.1 Water update properties

In the Kraton/GO/Ag/Pani composite-based IPMC soft actuator, the deformation of conventional IMPC is believed to be due to the movement of cations along with the water molecules under applied potential. Thus the higher WU is found to be responsible for the lower brittleness and better actuation performance of the IPMC membrane. The WU capacity of Kraton/GO/Ag/Pani membrane increases with increasing immersion time. Table 2.1 shows maximum WU capacities of the IPMC membrane at room temperature (25°C ± 3°C); for 10 h of immersion time it was found to be 314% and after that saturation occurred. However, there was a significant increase in the WU at higher temperature (80°C) and maximum water sorption (349%) occurred just after 2 h of immersion (Table 2.1). The high WU of Kraton/GO/Ag/Pani membrane supported the increasing hydrophilic nature and ultimately resulted in better performance, even at high temperatures under an applied potential. Water absorption is an important consideration for electrosensitive actuators because it determines the diffusion coefficients for the

Table 2.1 IEC, WU and PC of Kraton/GO/Ag/Pani composite IPMC.

Sample	WU% at R.T	WU% at 80°C	IEC (meq/g)	PC (S/cm)
Kraton/GO/Ag/Pani	280 (6 h)	270 (1 h)	1.88	1.302×10^{-3} (at 50 KHz)
Kraton/GO/Ag/Pani	314 (10 h)	349 (2 h)	—	8.417×10^{-4} (at 75 KHz)
Kraton/GO/Ag/Pani	314 (12 h)	349 (4 h)	—	7.507×10^{-4} (at 100 KHz)

encapsulated cations, swollen matrix volume, and the dielectric permeability of the matrix in response to applied potential [14]. The IEC of the Kraton/GO/Ag/Pani composite polymer membrane was found to be 1.88 meq/g of dry membranes. Higher IEC means that more immovable ionic groups (sulfonic acid group) are present in the polymer membrane which can exchange a large number of Na metal ions with the protons, being polar in nature, and absorb a large number of water molecules. The IEC reflects ion migration that is governed by the ionization activation energy between membrane actuator in response to applied electrical voltage [65]. The high value of IEC is also responsible for the deep insertion of metal particles onto the ion exchange membrane surfaces. The large numbers of metal nanoparticles on the IPMC membrane surfaces reduced the resistance which is necessary to enable the large tip displacements.

Impedance spectroscopy was used to evaluate the PC of Kraton/GO/Ag/Pani IPMC membrane. Fig. 2.2 depicts the impedance curve of the Z'' axis with Z' over a frequency range of 50, 75, and 100 kHz, which makes semicircles for the Kraton/GO/Ag/Pani IPMC membrane and behaves like an RC circuit. From these results, it is clear that the sample does not behave as pure resistance and has some capacitance characteristics [66]. This might be due to the addition of Ag, GO, and Pani, which develop the homogeneous medium in the hydrophilic IPMC membrane for the transportation of protons. The high-frequency region intercept of the semicircles in the impedance spectra shows that the intrinsic resistance of IPMC materials, the resistance offered by the electrolytic solution, and

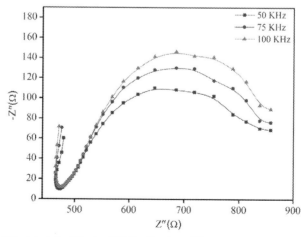

Figure 2.2 Nyquist plot of Kraton/GO/Ag/Pani at different scan rates.

the contact resistance of the interface decreases with the increase in the frequency range from 50–100 kHz. The maximum PC of Kraton/GO/Ag/Pani IPMC was found to be 1.302×10^{-3} S/cm under a frequency range of 50 kHz (Table 2.1). The actuation performance of the IPMC membrane is based on the fact that cations move quickly toward the cathode to create an imbalance pressure inside the IPMC actuator. Hence, the high PC of the IPMC membrane indicated that more hydrated cations are moved quickly toward the cathode side, showing a large displacement and fast actuation. Thus the rate of actuation of the IPMC membrane is increased with the enhancement of PC.

Fig. 2.3 represents the water loss of IPMC membrane examined as a function of time at room temperature. The maximum decrease in the mass loss of fully hydrated membrane reached 55% at the time of 3200 s. This loss of solvent from the IPMC may be attributed to the surface morphology and the fabrication process of the membrane and natural evaporation. In the beginning of Fig. 2.2, the water loss represents rapid weight loss and it then occurs much more slowly. This indicates that the change of weight of the IPMC membrane complies with exponential decay law [65]. It is noteworthy that the addition of Ag metal and GO to the membrane changes the solvent-holding capacity of Kraton and so the water-holding capacity of IPMC is smaller than the bare Kraton, defined by the porosity and density of metals and GO.

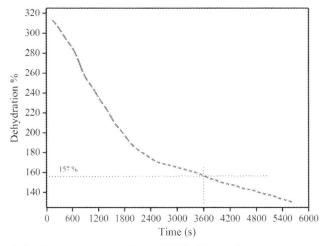

Figure 2.3 Dehydration process in Kraton/GO/Ag/Pani soft actuator.

2.4.1.2 FT-IR

FT-IR study was carried out in order to characterize the presence of GO, Ag nanopowder, and Pani composite in the Kraton polymer membrane. FT-IR spectra were recorded for pure Kraton membrane and Kraton/GO/AG/Pani composite-based IPMC actuator. Fig. 2.4A, the bands at ~3600 cm^{-1} is assigned to sulfonic acid groups O—H vibration, other bands were observed at 1078, 1023, and 735 cm^{-1} due to O=S=O, S=O and S—O stretches respectively, which are confirming the existence of block styrene sulfonate unit of Kraton polymer [67]. The peaks around 1600 and 1450 cm^{-1} ascribed due to the C=C stretching vibration of Kraton pentablock copolymer. The characteristic peak at ~3434 cm^{-1} attributed due to N—H stretching vibration of Pani (Fig. 2.4B). The bands at 1489 and 1571 cm^{-1} are due to the C=C stretching mode of the benzenoid and quinoid rings, respectively, and the peak at about ~1285 cm^{-1} can be assigned to C—N stretching. In Fig. 2.4B the peaks at 3373, 1718, 1591, and 1055 cm^{-1} are ascribed due to the OH, C=O, C—O in COOH and C—O in COH/COC (epoxy) functional groups, respectively, for GO [68]. From FTIR analysis, it is observed that after adding Go/Ag nanopowder and Pani the intensity of the peaks of the Kraton polymer reduces.

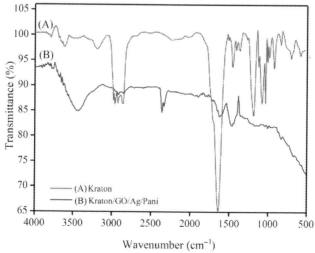

Figure 2.4 FT-IR spectra of Kraton and Kraton/GO/Ag/Pani.

2.4.1.3 SEM

The corresponding SEM images are used to check the distribution of Ag nanoparticles, dispersion of GO, and Pani coating in the porous surface of the Kraton ionic polymer membrane. The surface and cross-sectional morphologies of the Kraton/GO/Ag and Kraton/GO/Ag/Pani membranes actuator are shown in Fig. 2.5. It is clearly observed from the Fig. 2.5A and B that smooth and compact structures with densely deposited silver nanoparticles are wrapped within GO embedded on the surface and insides of pores of the hydrophilic porous surface of IPMC membrane. The silver nanoparticle along with GO can be clearly shown

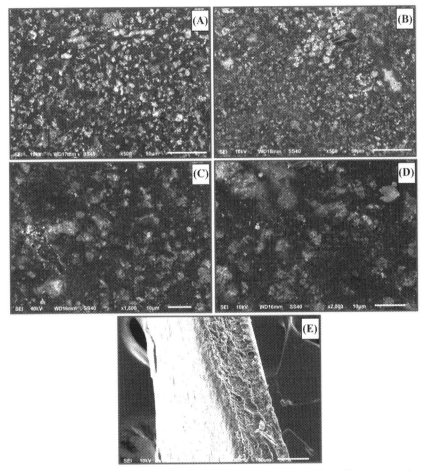

Figure 2.5 SEM Images of (A, B) Kraton/GO/Ag, (C, D) Kraton/GO/Ag/Pani composite-based actuator, and (E) cross sectional image showing coating of Pani.

predominately deposited all over the surfaces of the proposed IPMC actuator. The porous surface along with the Ag nanoparticles and GO in the Kraton/GO/Ag/Pani composite IPMC membrane, are important to enhance the transportation of ions in hydrophilic membrane actuator. The ease of transportation of ion increase the PC of the composite membrane, which is responsible for the large and fast actuation of the IPMC membrane. In addition, some surface roughness is observed on the Pani-deposited polymer membrane surface as a result of the cracking of the Pani layer during membrane drying before SEM analysis (Fig. 2.5C and D). Deposition of Pani film at the Kraton/GO/Ag membrane can be clearly seen in the cross section micrograph (Fig. 2.5E). Fig. 2.5D confirms there is no significant difference in the surface morphology after applied electrode potential, hence the main cause of water loss may be due to natural evaporation and electrolysis.

2.4.1.4 AFM
Fig. 2.6A and B show the three-dimensional and amplitude scan AFM images of the top surface, which features a modular structure with interconnected cavities. However, after the polymerization of Pani, as discussed earlier, the cavities and roughened surface were changed into a somewhat smooth and cavity-free surface with negligible surface cracks due to the deposition of the Pani layer (Fig. 2.6C and D). The three dimensional AFM image of the top surface of the Kraton/GO/Ag/Pani-based IPMC confirmed the cavity free, smooth surface morphology after deposition of conducting polymer layer. The surface morphology played an important role in the transport of cations and therefore in the bending behavior of the IPMC actuator. The rough and ruptured surface at some places is attributed to the surface roughening of the composite polymer membrane with abrasive paper before oxidative polymerization of Pani at the membrane surface. Fig. 2.6E and F shows that there is no significant change in the surface morphology after the applied electric potential.

2.4.1.5 UV–visible studies
The UV–visible spectra of the Kraton/GO/Ag/Pani IPMC membrane are used to determine the chemical interaction between the Kraton polymer membrane, GO, Ag nanoparticles, and Pani. The UV–visible absorbance pattern of Kraton, Kraton/GO/Ag, and Kraton/GO/Ag/Pani are shown in Fig. 2.7. Fig. 2.7A shows broad absorbance peaks at 203.99, 231.98, and 235.98 nm, which may be attributed to the characteristic

56 Soft Robotics in Rehabilitation

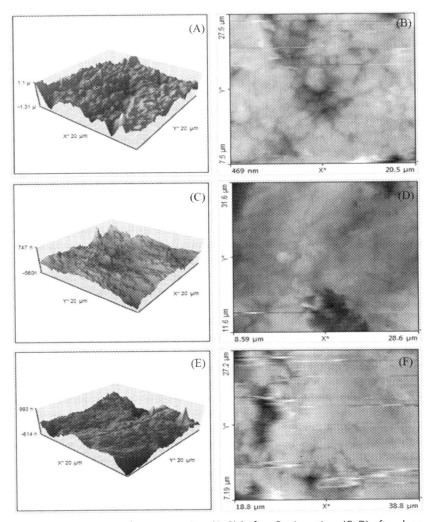

Figure 2.6 AFM images from a top view (A, B) before Pani coating, (C, D) after deposition of Pani layer, and (E, F) after actuation performance.

absorbance peaks of Kraton pentablock copolymer used as the base ion exchange material for the fabrication of the proposed IPMC actuator. Fig. 2.7B shows the absorbance pattern of Kraton/GO/Ag in which an additional absorbance peak is observed at 237.98 nm with the minor shifting of the second peak at the 235 nm wavelength, which proves that the interaction between Kraton, GO and Ag nanopowder. This indicates that the Ag nanoparticles and GO are embedded on the surface of the

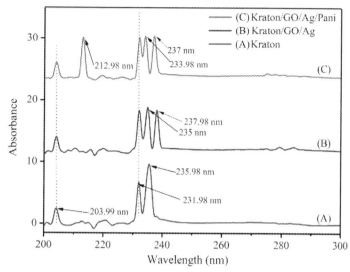

Figure 2.7 UV—visible pattern of (A) Kraton, (B) Kraton/GO/Ag, and (C) Kraton/GO/Ag/Pani.

composite membrane. In the Kraton/GO/Ag/Pani polymer composite, due to the deposition of Pani film on the Kraton/GO/Ag ionic polymer membrane, a sharp absorbance peak can be clearly seen at a shorter wavelength at 212.98 nm with the minor shifting of other peaks toward lower wavelengths (Fig. 2.7C). The additional absorbance peaks shown in Fig. 2.7C, other than seen in Fig. 2.7A, confirm the chemical interaction between Kraton, GO, Ag, and Pani-based composite used for the fabrication of IPMC actuator.

2.4.1.6 Electrochemical properties

The electrochemical performance of the Kraton/GO/Ag/Pani-based IPMC was analyzed using CV and LSV in 0.1 M sodium sulfate aqueous electrolyte solution at triangle voltage ±3 V under scan rate of 50—150 mV/s in a three-electrode system. All curves appearing to be similar in shape but differ in the magnitude of the current density (Fig. 2.8). As can be seen from LSV curve, the current density of IPMC at low scan rate shows the higher value of current density, which proves that the electrical current directly depends on the scan rate in the ionomeric polymer under the same applied voltage. This indicates that current density can be seen as a function of scan rate. Hence, the rate of ionic transfer in the IPMC actuator is larger at a lower scan rate, which is decreased by a

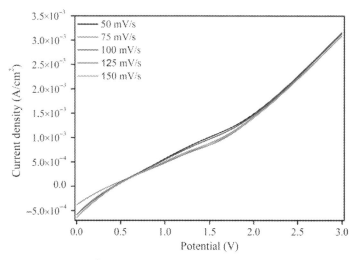

Figure 2.8 LSV curves of Kraton/GO/Ag/Pani in 0.1 M sodium sulfate solution at room temperature with a triangle potential ± 3 under scan rate of 50–150 mV/s.

higher scan rate. The symmetric shape of the CV curves can be assigned to excellent charge distribution in the whole surface region of the IPMC (Fig. 2.9). The peak current exhibited a linear increase with the square root of the scan rate (Fig. 2.10), indicating a diffusion-controlled redox process. Nevertheless, the Fig. 2.9 clearly shows that the I–V curve under scan rate of 50 mV/s has an irregular shape with different sequential small peaks. This means that the ionic diffusion in the IPMC membrane is chaotic and can be attributed to the rough and porous membrane surface. Furthermore, the electrode resistance increases at higher scan rate, hence, the anodic peak current shifted toward lower positive potential and cathodic current shifted in the reverse direction with the increase of scan rate (Fig. 2.9). It is interesting to note that the observed current density of the Kraton/GO/Ag/Pani membrane is remarkably higher in comparison to Kraton and several other IPMCs reported so far [64–66], which may be due to the addition and conductive Pani coating followed by the increased charge transfer due to the increase in surface area for chemical reactions by electrically conductive GO. The benefit of the increased conductivity and capacity due to Pani and GO is indicated by the more capacitor-like shape of the I versus E curve under scan rate of 50 mV/s. The higher current density which reflects the energy storage ability, hence, is ascribed to the better actuation performance of IPMC by larger deformation of polymer membrane under applied potential.

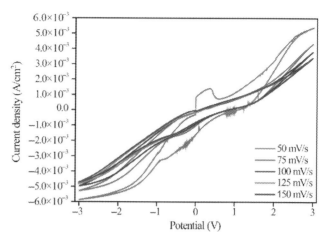

Figure 2.9 CV curves of Kraton/GO/Ag/Pani in 0.1 M sodium sulfate solution at room temperature with a triangle potential ± 3 under scan rate of 50–150 mV/s.

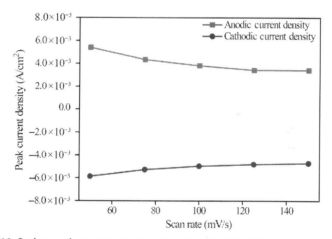

Figure 2.10 Redox peak current versus scan rate (50–150 mV/s).

The capacitance is important to reflect the electrical properties of IPMC membrane. A greater energy storage ability is implied by the high electric current, which determines the bending behavior of the membrane actuator. Fig. 2.11 shows that the capacitance values of the Kraton/GO/Ag/Pani-based IPMC are dependent on scan rate. The proposed IPMC membrane show the remarkable capacitance values, especially 222.56 µF under scan rate 50 mV/s in the fully hydrated state. Layer by layer coating

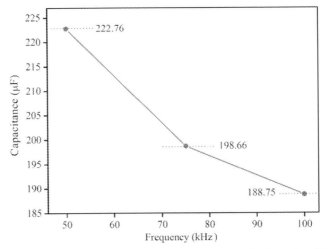

Figure 2.11 Capacitance behavior of Kraton/GO/Ag/Pani IPMC under different frequency range.

of Pani and uniformly distribution of Ag nanoparticles inside the ionomeric polymer membrane may be considered as the key factors for high capacitance value. Furthermore, decrease in the capacitance value is ascribed due to the decrease in the rate of ionic transfer as discussed before at higher scan rate.

2.4.1.7 Electromechanical properties

In order to examine the behavior of Kraton/GO/Ag/Pani-based IPMC actuator, an actual testing setup was developed, as shown in Fig. 2.12. An ionic actuator is clamped in a cantilever configuration in a holder and connected to digital power supply where the controlled voltage (± 3 V) is sent through the NI-PXI System. For controlling this voltage, a VI is developed in Labview software. The PID control system feature is also incorporated in the PID algorithm. A laser displacement sensor is also in front of an ionomeric actuator for measuring the displacement. This is also interfaced with the NI-PXI System through an RS-485 to the RS-232 convertor. The DC power (12 V) is sent to a laser displacement sensor through a variable voltage supply. The displacement data from the laser sensor is acquired into the NI-PXI through comport and NI Visa interfaces software module in a Labview VI. Simultaneously, the actuating voltage is obtained from the programmable power supply through a data acquisition (DAQ) assistant of NI-PXI system.

Development of different types of ionic polymer metal composite-based soft actuators   61

Figure 2.12 Experimental testing setup for the Kraton/GO/Ag/Pani-based IPMC actuator.

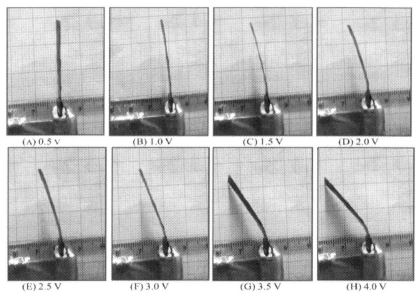

Figure 2.13 Successive steps for activating an IPMC actuator. (A) Bending of the soft actuator at 0.5 V, (B) Bending of the soft actuator at 1.0 V, (C) Bending of the soft actuator at 1.5 V, (D) Bending of the soft actuator at 2.0 V, (E) Bending of the soft actuator at 2.5 V, (F) Bending of the soft actuator at 3.0 V, (G) Bending of the soft actuator at 3.5 V, (H) Bending of the soft actuator at 4.0 V.

For characterization of IPMC, the sinusoidal voltage (± 4 V) is sent to the Kraton/GO/Ag/Pani-based IPMC actuator. The successive steps of bending response are shown in Fig. 2.13. The maximum deflection of proposed actuator is shown to be up to 18 mm. The different trials were

Table 2.2 Experimental deflection data of Kraton/GO/Ag/Pani IPMC actuator.

Deflection (mm)	Voltage (V)								
	0 V	0.5 V	1.0 V	1.5 V	2.0 V	2.5 V	3.0 V	3.5 V	4.0 V
d1	0	1.0	4.0	6.0	8.0	9.0	11.0	15.0	18.0
d2	0	0.9	4.1	6.0	7.9	8.9	11.0	15.0	17.8
d3	0	0.6	3.9	5.8	8.0	9.0	10.8	14.8	18.0
d4	0	0.8	3.6	5.5	7.6	8.8	9.9	14.6	17.8
d5	0	0.9	3.9	5.8	7.9	8.5	9.6	14.5	17.5
d6	0	1.0	3.5	5.4	7.5	8.3	9.1	14.6	17.7
d7	0	0.7	3.7	5.8	7.1	8.6	9.5	14.2	17.5
d8	0	0.8	3.9	5.8	7.3	8.5	9.7	14.6	17.8
d9	0	0.7	3.7	5.5	7.6	8.4	9.2	14.4	17.6
d10	0	0.6	3.5	5.8	7.7	8.1	9.2	14.7	17.8

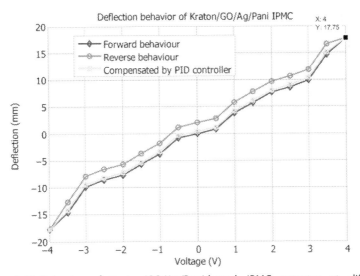

Figure 2.14 Behavior of Kraton/GO/Ag/Pani-based IPMC actuator at different voltages.

conducted and the deflection data were collected, as given in Table 2.2. The average value of deflection at different voltage is plotted as shown in Fig. 2.14. When a voltage is applied, it shows steady-state behavior but after reducing the voltage, it does not provide the same path and it has some deflection error. This is reduced by controlling the voltage through the PID controller system where PID parameters are tuned in Labview software. The deflection error is minimized up to 80%.

For load characterization of an ionic polymer actuator, the actuator was clamped in a cantilever configuration. A digital weighing/load cell (Model: Citizen CX-220, Make: India) was used where the tip of the ionic polymer actuator touched the pan of load cell during applied voltage through the NI-PXI System (Fig. 2.15). The Kraton/GO/Ag/Pani-based IPMC actuator bends and the maximum load carrying capability was up to 51 mg at 4 VDC. To find the repeatability of this actuator, several trials were conducted and the force data were noted for 10 times, as given in Table 2.3. After taking an average of 10 force values, the mean force value was obtained and the normal distribution was calculated, as given in Table 2.3. The normal distribution was drawn as shown in Fig. 2.16. The normal distribution is very narrow, which shows less error and great repeatability of the forced behavior at a particular voltage (4.0 V). The comparison of different properties with different kinds of IPMCs is summarized in Table 2.4. It shows that the properties of the Kraton/GO/Ag/Pani-based IPMC is much better than other types of IPMCs based on expensive Pt metal as an electrode material.

2.4.2 Characterization of SPVA/IL/Pt IPMC composite-based IPMC soft actuator

The following characterizations of SPVA/IL/Pt IPMC composite-based IPMC soft actuator were carried out.

2.4.2.1 WH, IEC, and PC

The types of composite blends, nature of the polymer, fixed ionic groups, counterion, and water-holding capacity are the key factors which affect

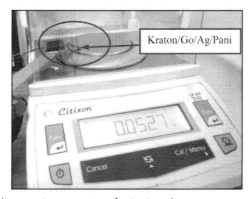

Figure 2.15 Load measuring test setup for ionic polymer actuator.

Table 2.3 Experimental force data of Kraton/GO/AG/Pani-based IPMC actuator.

Force (mN)	Voltage (V)				
	0 V	1 V	2 V	3 V	4 V
F1	0	0.0150	0.0217	0.0443	0.0527
F2	0	0.0150	0.0217	0.0443	0.0525
F3	0	0.0148	0.0216	0.0441	0.0526
F4	0	0.0146	0.0214	0.0438	0.0524
F5	0	0.0143	0.0211	0.0437	0.0521
F6	0	0.0141	0.0211	0.0435	0.0522
F7	0	0.0137	0.0208	0.0433	0.0520
F8	0	0.0132	0.0205	0.0430	0.0518
F9	0	0.0129	0.0201	0.0428	0.0515
F10	0	0.0124	0.0200	0.0427	0.0514
Operating voltage					4.0 V
Mean					0.0521 mN
Standard deviation					0.0449 mN
Repeatability					95.51%

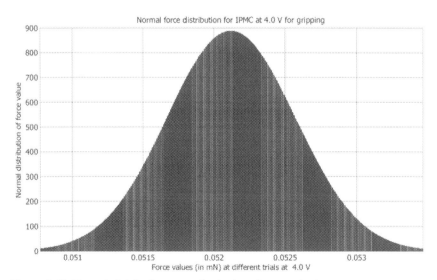

Figure 2.16 Normal distribution curve for ionic polymer actuator.

the performance of IPMC actuators. It was found that the PVA membranes after sulfonation were unstable in water. Therefore water-holding capacity (WH) of the SPVA membranes cannot be measured accurately

Table 2.4 Comparison of Kraton/GO/AG/Pani-based IPMC with other IPMC actuators.

Parameter	Kraton/GO/Ag/Pani	Nafion [69]	Sulfonated polyetherimide [70]	Carbon nanotube actuator [71]	Sulfonated polyvinyl alcohol/Py [72]
Tip displacement (mm)	17.5	12	2.7	20	18.5
Ion-exchange capacity (meq/g)	1.88	0.98	0.553	0.71	1.2
Water uptake (%)	314	16.70	26.4	25.1	82.3
Current density (A/cm^2)	0.0054	0.03	5×10^{-4}	—	0.0055
Proton conductivity (S/cm)	1.302×10^{-3}	9.0×10^{-3}	1.4×10^{-3}	5.7×10^{-3}	1.6×10^{-3}

because of the partial dissolution of membranes in DMW. This effect might be attributed to a relatively high polarity of the SPVA membrane. To attain the stability in DMW or WH to an adequate amount to avoid the disadvantage of partial dissolution of fabricated ionomeric polymer membranes, it was suggested that the cross-linking process might be carried. Therefore the SPVA/IL membranes were treated at a temperature of 120°C for 1 h to carry out cross-linking. After cross-linking the prepared polymer membrane was kept in DMW at 60°C for 24 h. The results show that the mass of SPVA/IL membrane almost remained constant after extracted by DMW, suggesting that the SPVA membrane was crosslinked and there was no free SPVA chain in the resulting membrane. Although due to the presence of ionic liquid the developed IPMC was capable of working in the dry condition. Besides this, the water-holding capacity (WH) of SPVA/IL/Pt IPMC was calculated at room temperature and 60°C. The water-holding capacity of SPVA/IL/Pt membrane at room temperature (25°C ± 3°C) and 60°C was found to be a maximum of 140% ± 0.24% and 190% ± 0.40% after 10 h of immersion, respectively, after that saturation was found to be predominant (Table 2.5). The higher water-holding capacity of SVPA/IL/Pt IPMC membrane may be due to the presence of more active thermally enlarged $-SO_3H$ sites. There was no breakage and surface rupturing in fully hydrated IPMC membrane even at increasing temperature ($\geq 60°C$), which may enable the movement of more hydrated cations [73]. To check the stability of SPVA/IL/Pt IPMC membrane, the WH testing at RT and 60°C for a different interval of times was repeated four times. The average of these four repeated experiments was taken to plot an error bar of WH (Fig. 2.17) for SPVA/IL/Pt membrane. This analysis suggested that there was no significant change in the WH capacity of the proposed IPMC after repetition of the same experiment. This, in turn, confirms the stability of SPVA/IL/Pt IPMC toward solvent-holding capability. The elevated WH of SPVA/IL/Pt IPMC supported the hydrophilic nature and ultimately confirmed the better performance under an applied potential [66]. IEC strongly influences the cations transfer through the ion exchange membrane and determines the better performance of the IPMC actuator. The IEC of SPVA/IL/Pt-based IPMC membrane was found to be 2.1 ± 0.015 meq/g of the dry membrane, as shown in Table 2.5. The high IEC of the membrane allows more Pt particles to deeply embed on the surfaces of the SPVA/IL/Pt membrane. More Pt particles cause uniform electroding and low resistance, hence it is responsible for fast and

Table 2.5 IEC, DS, PC and WH of SPVA/IL/Pt-based IPMC actuator.

Materials	IEC (meq/g of dry membrane)	DS (%)	PC (σ) (S/cm)	WH (%) at R.T	WH (%) at 60°C
SPVA/IL/Pt	2.1 ± 0.015	26.22 ± 0.372	$1.45 \times 10^{-3} \pm 1.78 \times 10^{-5}$	140 ± 0.24 (10 h)	190 ± 0.40 (10 h)
SPVA/IL/Pt	—	—	—	140 ± 0.31 (20 h)	190 ± 0.43 (20 h)

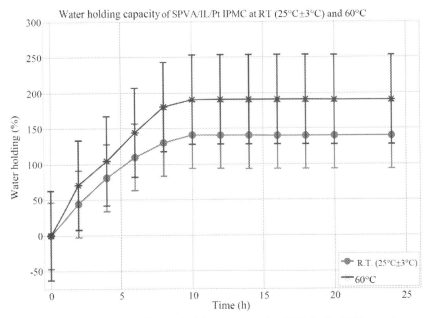

Figure 2.17 Error bar plot of water-holding capacity for SPVA/IL/Pt IPMC membrane.

large bending performance [67]. From the results of IEC, the DS of the proposed IPMC membrane was determined to be 26.22% ± 0.372% (Table 2.5), which is adequately high for the performance of the IPMC actuator. The PC of the SPVA/IL/Pt-based IPMC actuator was found to be 1.45×10^{-3} S/cm with a standard deviation (SD) of $\pm 1.78 \times 10^{-5}$ S/cm (Table 2.5). High PC enables the movement of more metal cations in their hydrated state, which results in fast and large actuation [66].

2.4.2.2 Water loss

The main cause of performance decay and short lifetime of IPMC is the inner solvent loss from the damage of the membrane surface [66]. Natural evaporation, electrolysis, and leakage from the damaged or porous surface are the main causes of solvent or water loss from the membrane actuator. Water loss experiment of SPVA/IL/Pt IPMC membrane at RT for a time interval of 2, 4, 6, 8, and 10 min under applied voltage of 3 V was repeated five times. The maximum water loss for SPVA/IL/Pt IPMC membrane actuator was found to be 52% ± 0.5016% under an applied potential of 3 V for a maximum time of 10 min (Fig. 2.18). The results of water loss suggested that there was no considerable change in the water

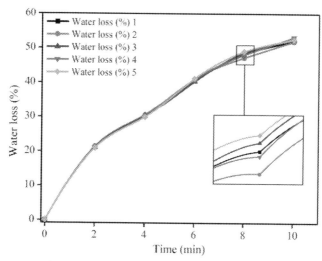

Figure 2.18 Water loss (%) of SPVA/IL/Pt-based IPMC actuator.

loss of the IPMC membrane after repetition of the same experiment. This, in turn, confirms the main cause of water loss was natural evaporation and electrolysis because there was no significant difference in the surface morphology after applied electrical potential [68].

2.4.2.3 SEM and EDX

Surface morphology is responsible for the better performance of an IPMC actuator under applied electrical potential. Fig. 2.19A and B shows the surface morphology of the SPVA/IL/Pt-based IPMC membrane before and after applied electrical potential, respectively. The smooth and almost cavity-free surface of the SPVA/IL/Pt membrane (Fig. 2.19A) after testing actuation changes into the slightly rough surface with a few ruptures after applied electric voltage (Fig. 2.19B). Therefore the surface morphology of the membrane after applied electrical potential showed no significant changes. Hence, the electrolysis and natural water evaporation were the main causes for the solvent loss for the fabricated IPMC membrane. In Fig. 2.19C the cross-sectional image of the SPVA/IL/Pt membrane clearly shows the Pt electrode layer on the surface of the IPMC membrane.

Fig. 2.20 shows the EDX results for the fabricated IPMC membrane. The EDX spectrum shows that the surface of the SPVA/IL/Pt IPMC membrane have the characteristic peaks of carbon (C), oxygen (O), sulfur

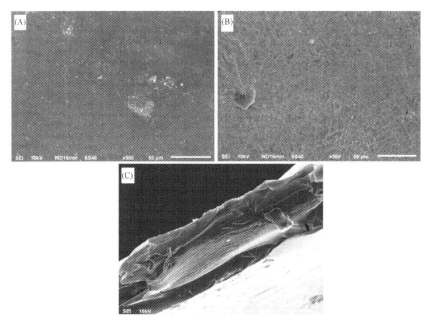

Figure 2.19 SEM micrographs of SPVA/IL/Pt IPMC membrane (A) before actuation, (B) after actuation, and (C) cross-sectional view.

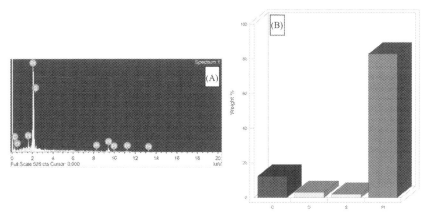

Figure 2.20 EDX spectra of SPVA/IL/Pt-based IPMC membrane. (A) Characteristics Peaks of Carbon (C), Oxygen (O), Sulfur (S) and Platinum (Pt), (B) Percentages of said elements on the surface of the soft actuator.

(S), and platinum (Pt) (Fig. 2.20A), while the percentages of these elements on the surface of the fabricated IPMC membrane are shown in Fig. 2.20B. The sulfonation of PVA was also confirmed by the presence of sulfur (S). The characteristic peak of Pt confirms the coating of the metal electrode on the surface of the SPVA/IL/Pt IPMC membrane.

2.4.2.4 Fourier-transform infrared and XRD study

Fig. 2.21 shows the FTIR spectrum of the SPVA/IL/Pt-based IPMC actuator. Two broad bands were observed at 3400 and 2950 cm^{-1}; the former peak was assigned to the O—H stretching mode in the hydroxyl groups (—OH), and the latter was assigned to the asymmetric —CH$_2$ stretching band. The bands appearing between 1000 and 1150 cm^{-1} were assigned to C—O stretching [64]. Further, characteristic bands appearing at approximately 1266 and 1750 cm^{-1} were assigned to the asymmetric stretching vibrations of S=O in the sulfonic acid group and the C=O stretching peak of carboxyl groups, respectively [70].

Representative XRD pattern of SPVA/IL/Pt IPMC membrane indicates little peaks of 2θ values as shown in Fig. 2.22. Generally, the diffraction patterns are sharp with high intensities of peaks, when there are crystalline domains, whereas the diffraction patterns are broad for the amorphous nature of the material. The XRD pattern of SPVA/IL/Pt exhibits a typical peak at $2\theta = 20.99$ degrees due to the presence of PVA [64]. Due to the sulfonation of PVA, the intensity of the diffraction pattern for PVA is decreased. Moreover, the XRD peaks obtained at $2\theta = 38.65$, 45.36, and 68.11 degrees correspond to the presence of Pt particles [71]. The presence of small peaks of 2θ values of Pt suggests that these particles were deeply embedded on the surface of the developed

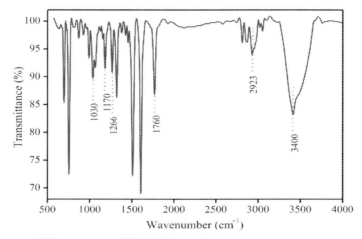

Figure 2.21 FT-IR spectrum of SPVA/IL/Pt membrane.

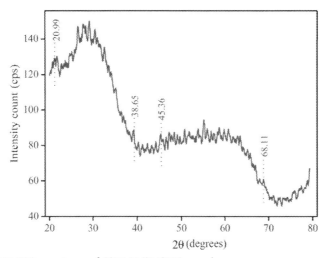

Figure 2.22 XRD spectrum of SPVA/IL/Pt IPMC membrane.

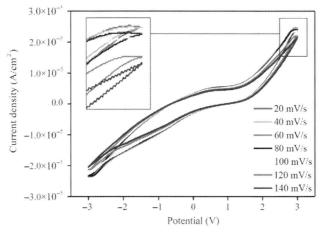

Figure 2.23 CV curve of SPVA/IL/Pt-based IPMC in 0.01 M NaOH at different scan rate.

IPMC membrane. The results of this XRD spectrum show an amorphous nature of the fabricated SPVA/IL/Pt-based IPMC membrane.

2.4.2.5 Electrical properties

The electrical performance of SPVA/IL/Pt IPMC membrane was confirmed by CV and LSV curves recorded under ± 3 V and 0–3 V, respectively, at scan rate 20–140 mV/s, as shown in Figs. 2.23 and 2.24. The impact of

Development of different types of ionic polymer metal composite-based soft actuators 73

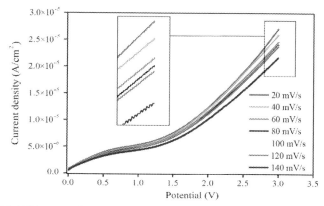

Figure 2.24 LSV curve of SPVA/IL/Pt-based IPMC in 0.01 M NaOH at different scan rate.

various scan rates on the current density of the fabricated IPMC membrane is clearly shown in Fig. 2.23. It is evident that the current density of the IPMC membrane decreases with the expansion in scan rates and all CV curves have similar shape but different magnitude of the current density. The LSV curve (Fig. 2.24) also shows a similar behavior and demonstrates the maximum current density at low scan rate. That shows the good electrocatalytic nature of the prepared IPMC membrane and proves that the electrical current directly depends on the scan rate in the ionomeric polymer under the same applied voltage. The results reveal that current density can be seen as a function of scan rate, that is, the rate of ionic transfer in IPMC membrane is larger at lower scan rate. The shape of current–voltage curve is reflected by movement of the hydrated metal cations under an applied voltage, with the decomposition profile of water due to electrolysis. It is clear from Fig. 2.10 that as the applied electric voltage increases from 1 to 3 V, the current density of SPVA/IL/Pt IPMC membrane was also increased accordingly. It means that the tip displacement of the SPVA/IL/Pt IPMC membrane has a proportional relationship with the dissipated power according to the increase of the driving voltages.

2.4.2.6 Electromechanical properties

For electromechanical characterizations, the voltage between 0 and 4 V DC was applied to SPVA/IL/Pt IPMC through a computer controlled DAC and a microcontroller. The voltage of 9−12 V DC was given to a microcontroller where a custom amplifier circuit was used for providing the desired current rating (50−200 mA) to the SPVA/IL/Pt IPMC membrane actuator. For proper connectivity between the conductive layers

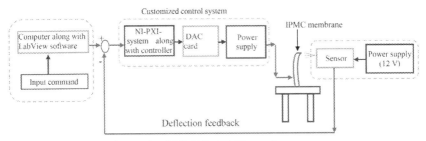

Figure 2.25 Control scheme for SPVA/IL/Pt IPMC membrane.

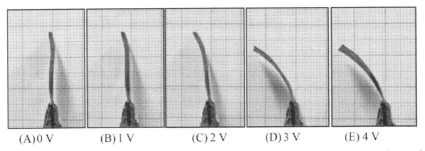

Figure 2.26 Stepwise deflection of SPVA/IL/Pt IPMC membrane. (A) Bending of SPVA/IL/Pt IPMC actuator at 0 V, (B) Bending of SPVA/IL/Pt IPMC actuator at 1 V, (C) Bending of SPVA/IL/Pt IPMC actuator at 2 V, (D) Bending of SPVA/IL/Pt IPMC actuator at 3 V, (E) Bending of SPVA/IL/Pt IPMC actuator at 4 V.

and wires, a copper tape was used to connect both surfaces of SPVA/IL/Pt membrane to this custom amplifier circuit. The tip displacement of the SPVA/IL/Pt membrane was measured using a displacement sensor under applied voltages. This was also used as a displacement feedback for controlling and measuring the tip displacement, where date conversion took it by converting the data from RS-485 to RS-232 communication protocol in order to attain the proper communication between sensor and input command from a computer with Docklight V1.8 software. A computer code was written in C programming language where the sampling rate (20 samples per second) was set for observing the bending of the membrane actuator.

After applying the voltage, the IPMC membrane bends and provides the dexterous behavior for manipulation. The basic control scheme is shown in Fig. 2.25. The successive steps of deflection of developed IPMC membrane are obtained under the applied voltage (0–4 V DC) as shown in Fig. 2.26. The repeatability of the fabricated IPMC membrane was

Development of different types of ionic polymer metal composite-based soft actuators 75

Table 2.6 Experimental data of deflection for SPVA/IL/Pt membrane actuator.

Voltage (V)	Deflection data (mm)					Average value of deflection (mm)
	D1	D2	D3	D4	D5	
0	0	0	0	0	0	0
1.0	2.20	2.30	2.10	1.90	1.90	2.08
2.0	4.50	4.20	3.80	3.90	3.90	4.06
3.0	12.20	11.50	12.00	11.90	12.10	11.94
4.0	14.20	14.00	13.50	13.80	13.90	13.88
Standard deviation at 4 V						± 0.232

Table 2.7 Deflection data of SPVA/IL/Pt IPMC membrane for trial 2.

Voltage (V)	Deflection data (mm)					Average value of deflection (mm)
	D1	D2	D3	D4	D5	
0	0.026	0.389	0.713	0.365	0.624	0.424
1.0	2.250	3.006	2.560	2.303	2.388	2.501
2.0	5.884	6.154	8.433	7.856	9.682	7.602
3.0	12.294	11.581	12.085	11.953	14.064	12.396
4.0	13.369	13.088	13.671	13.878	14.073	13.616
Standard deviation at 4 V						± 0.3938 mm

Table 2.8 Deflection data of SPVA/IL/Pt IPMC membrane for trial 3.

Voltage (V)	Deflection data (mm)					Average value of deflection (mm)
	D1	D2	D3	D4	D5	
0	0.185	0.635	0.619	0.479	0.002	0.384
1.0	2.489	2.988	2.418	2.188	2.561	2.529
2.0	5.473	6.4862	7.727	7.299	5.034	6.404
3.0	12.317	11.561	12.170	11.914	13.989	12.381
4.0	14.340	14.014	13.591	13.938	14.027	13.982
Standard deviation at 4 V						± 0.2674 mm

checked by conducting the deflection experiment for 4 × 5 (20) times. The deflection data for the first trial is provided in Tables 2.6–2.9. The maximum tip deflection of the fabricated IPMC actuator was observed up to 14.2 mm with the SD of ± 0.232 mm at 4 V DC. To observe the

Table 2.9 Deflection data of SPVA/IL/Pt IPMC membrane for trial 4.

Voltage (V)	Deflection data (mm)					Average value of deflection (mm)
	D1	D2	D3	D4	D5	
0	0.583	0.334	0.065	0.408	0.135	0.305
1.0	2.585	2.557	2.128	2.426	2.453	2.430
2.0	7.667	7.019	7.517	8.110	8.215	7.684
3.0	12.319	11.666	12.016	12.059	13.989	12.410
4.0	14.307	14.055	13.513	13.949	13.925	13.950
Standard deviation at 4 V						± 0.2873 mm

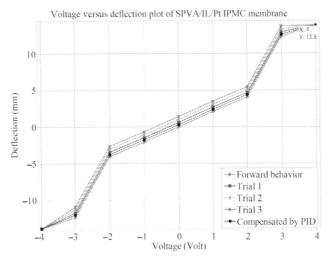

Figure 2.27 Bending behavior of SPVA/IL/Pt IPMC actuator over applied voltage.

hysteresis behaviors at multiple repeats (20 times), the average value of each five experiments was taken as one trial. Therefore different trials were plotted as shown in Fig. 2.27. The results show that, when the voltage was increased, SPVA/IL/Pt IPMC attempts the linear bending behavior but after reducing the voltage from high to low the IPMC did not appear to use the same path to return to its original position. A PID control system was used for minimizing the deflection error of SPVA/IL/Pt IPMC membrane, where the controller bandwidth was set by tuning the frequency. When the frequency in the controller was increased, the time duration of actuation of SPVA/IL/Pt-based IPMC decreased but at the

Table 2.10 Force data of SPVA/IL/Pt-based IPMC actuator.

Voltage (V)	f_1 N ($\times 10^{-2}$)	f_2 N ($\times 10^{-2}$)	f_3 N ($\times 10^{-2}$)	f_4 N ($\times 10^{-2}$)	f_5 N ($\times 10^{-2}$)	Average force value (f) in N ($\times 10^{-2}$)
0	0	0	0	0	0	0
1.0	0.0145	0.0146	0.0129	0.0138	0.0137	0.0139
2.0	0.0302	0.0322	0.0308	0.0308	0.0330	0.0309
3.0	0.0412	0.0412	0.0395	0.0405	0.0440	0.0406
4.0	0.0448	0.0449	0.0454	0.0438	0.0443	0.0444
Mean						0.0324
Standard deviation						± 0.00049
Repeatability						87.2%

same time, the stability of the IPMC also decreased. In order to achieve the fast response of IPMC, by proper tuning of gains in the PID controller the frequency was set. This reveals the steady position between the response time and stability. This customized system, using a PID controller in the NI-PXI system was applied for controlling the IPMC link manipulator because it provides minimization with the error and also reduces overshoot (error on the other side because of too great applied force) during manipulation of flexible components. After that, this ionic actuator can be used for dexterous handling in manipulation (Fig. 2.27).

To determine the force behavior the SPVA/IL/Pt-based IPMC membrane was clamped into a holding fixture in the horizontal position. A multimeter was used for measuring the voltage simultaneously while the membrane actuator was in operation. The five different force values (f_1, f_2, f_3, f_4, and f_5) were collected at different voltage ranges (0–4 V), as given in Table 2.10. After that, the mean force values were calculated from five different readings. From the corresponding mean values, the SD was obtained. To determine the mechanical properties, the stress behavior with respect to the applied potential was envisaged, as shown in Fig. 2.28. When the applied voltage increases, the stress behavior of the developed IPMC actuator is also increased continuously up to 4 V DC. The maximum generated stress for SPVA/IL/Pt IPMC membrane at 4 V DC was found to be 15.01 Pa with the SD of ± 0.1626. Fig. 2.29 shows the normal distribution function of SPVA/IL/Pt IPMC actuator. By using the normal distribution function, the repeatability of the SPVA/IL/Pt IPMC actuator was found to be 87.2%. A comparison of SPVA/IL/Pt-based IPMC actuator with other reported actuators is given in Table 2.11.

78 Soft Robotics in Rehabilitation

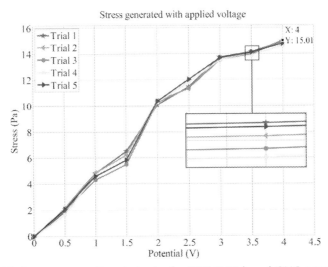

Figure 2.28 Stress versus voltage behavior for SPVA/IL/Pt-based IPMC membrane.

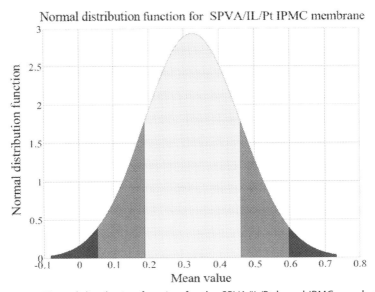

Figure 2.29 Normal distribution function for the SPVA/IL/Pt-based IPMC membrane.

Table 2.11 Comparison of SPVA/IL/Pt-based IPMC actuator with other reported actuators.

Parameters	SPVA/IL/Pt	Nafion [70]	Polyacrylonitrile–Kraton-GO [68]	Kraton-based IPMC [66]	Sulfonated polyetherimide [64]	CNT-based actuator [71]
WU (%)	190	16.70	133	233	26.4	25.1
Current density (A/cm^2)	3.0×10^{-3}	3×10^{-3}	1.5×10^{-4}	2.5×10^{-3}	5×10^{-4}	–
Tip displacement (mm)	14	12	16	17	2.7	20
IEC (meq/g^1)	2.1	0.98	1.4	2	0.553	0.71
PC (S/cm^1)	1.45×10^{-3}	9.0×10^{-2}	5.26×10^{-3}	1.302×10^{-3}	1.4×10^{-3}	5.7×10^{-3}

2.5 Development of robotic system using different types of IPMC actuators

2.5.1 The design concept of the flexible link manipulator using SPVA/IL/Pt IPMCs for robotics assembly

In order to develop the flexible link manipulator, a novel design concept was developed, as shown in Fig. 2.30. The SPVA/IL/Pt IPMC-based flexible link (size 40 mm × 10 mm × 0.2 mm) was integrated between plastic-based links with similar in size. This IPMC was used as an active flexible joints link during manipulation and the same IPMCs were also used as fingers during the development of a micro gripper. This micro gripper was integrated at one end of the plastic link for holding the object. The IPMC links were activated by applying the voltages (± 4 V) independently. These voltages were controlled through the NI-PXI controller. The major advantages of this IPMC in the assembly were that the SPVA/IL/Pt IPMCs were compliant in nature and provided more flexibility. The developed IPMCs-based flexible manipulator can hold the different types of object like rectangles, circles, pentagons, etc.

A novel flexible link manipulator using the SPVA/IL/Pt-based IPMC membrane was developed, as shown in Fig. 2.31. An IPMC was considered as a flexible joint link and two IPMC fingers-based micro gripper were also integrated at the end of the flexible link. The flexible IPMC

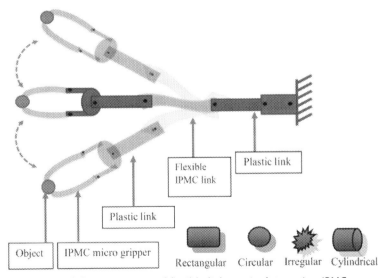

Figure 2.30 Novel design concept of flexible link manipulator using IPMCs.

Development of different types of ionic polymer metal composite-based soft actuators   81

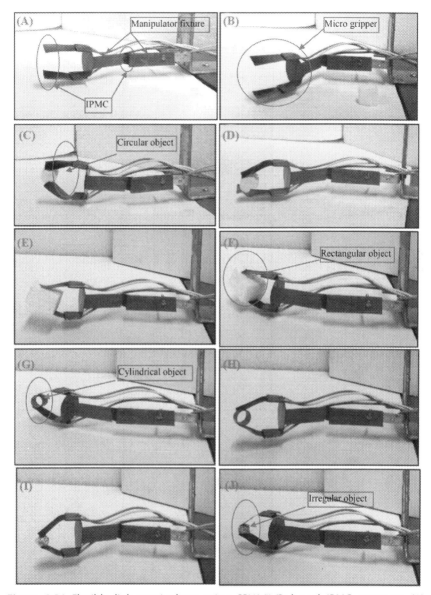

Figure 2.31 Flexible link manipulator using SPVA/IL/Pt-based IPMC actuators. (A) Actuation of SPVA/IL/Pt IPMC actuator at Joint, (B) Actuation of SPVA/IL/Pt IPMC actuators of microgripper, (C) Holding of a circular object by microgripper, (D) Holding of irregular shape by microgripper, (E) Holding square shape object by microgripper (F) Holding of a rectangular object by microgripper, (G) Holding of cylindrical ring object by microgripper, (H) Manipulation of the object, (I) Holding of a solid irregular object by microgripper, (J) Manipulation of an irregular object.

Table 2.12 Capability of manipulation of different type of objects.

S. no.	Object	Weight (mg)
1.	Circular object	21.1
2.	Rectangular object	36.6
3.	Irregular shape object	76.6
4.	Cylindrical object	117.0

joint provides the bidirectional bending of the link manipulator for manipulating objects of different shapes and sizes from one position to another position. The IPMC micro gripper holds the object by providing the applied voltage (0−4 VDC). This shows the potential of dexterous handling of the object. The handling capability of the flexible manipulator using SPVA/IL/Pt IPMC is demonstrated which can hold the different types of object as given in Table 2.12. The weight of objects with different shapes and sizes that can be held by the developed manipulator ranged from 20 to 117 mg. During assembly it was also demonstrated that this kind of manipulator provides a more flexible nature compared with as rigid type manipulator because in this manipulator the IPMC was used as an active joint during manipulation. In addition, the IPMCs-based micro gripper was also used for handling objects with any kind of shape. These were the major advantages of this kind of flexible manipulator for robotic assembly.

2.6 Conclusion

This chapter describes the development of the novel electrochemically driven Kraton/GO/Ag/Pani composite-based soft actuator and an efficient nonperfluorinated SPVA/IL/Pt-based IPMC actuators using a solution casting method. By characterizing the Kraton/GO/Ag/Pani composite and SPVA/IL/Pt IPMC, it were found that these can be operated in dry and hydrated states. The higher water-holding capacity, PC, high IEC, and slow water loss after applying the electric potential provide the better performance and repeatability of the SPVA/IL/Pt-based IPMC actuator as compared to the other IPMCs. The electrical properties were also confirmed by CV and LSV, which support the better actuation performance of the fabricated different IPMC-based soft actuators. By developing a prototype link manipulator, it was demonstrated that Kraton/GO/Ag/Pani composite and SPVA/IL/Pt-based IPMC soft actuators have more flexible and compliant behavior during the manipulating and handling of objects of different shapes and sizes. This potential proved that the fabricated Kraton/GO/Ag/Pani composite and SPVA/IL/Pt-based IPMC can be used as active and compliant fingers and joints in dexterous handling devices in the robotic assembly. In future, these novel soft actuators can substitute for rigid link mechanisms in the field of microrobotics applications, such as in micro grippers, bioengineering, biologically-inspired robotic systems, and other robotic applications.

Acknowledgment

The authors are thankful to the Director, CSIR-CMERI, India for granting permission to publish this book chapter paper.

References

[1] M. Shainpoor, Y. Bar-Cohen, J.O. Simpson, J. Smith, Ionic polymer-metal composites (IPMC) as biomimetic sensors, actuators, and artificial muscle — a review, Int. J. Smart Mater. Struct. 7 (1998) R15—R30.
[2] M. Shahinpoor, K.J. Kim, The effect of surface-electrode resistance on the performance of ionic polymer-metal composite (IPMC) artificial muscle, Smart Mater. Struct. 9 (4) (2000) 543—551.
[3] M. Shahinpoor, K.J. Kim, Ionic polymer-metal composites: I. Fundamentals, Smart Mater. Struct. 10 (2001) 819—833.
[4] S. Nemat-Nasser, Y. Wu, Comparative experimental study of ionic polymer-metal composites with different backbone ionomers and in various cation forms, J. Appl. Phys. 93 (9) (2003) 5255. Available from: https://doi.org/10.1063/1.1563300.
[5] R.K. Jain, S. Majumder, A. Dutta, SCARA based peg-in-hole assembly using compliant IPMC based micro gripper, Robot. Auton. Syst. 61 (3) (2013) 297—311.
[6] R.K. Jain, S. Datta, S. Majumder, A. Dutta, Two IPMC fingers based micro gripper for handling, Int. J. Adv. Robot. Syst. 8 (1) (2011) 1—9.
[7] R.K. Jain, S. Majumder, A. Dutta, Micro assembly by an IPMC based flexible 4-bar mechanism, Smart Mater. Struct. 21 (7) (2012). Available from: 10.1088/0964-1726/21/7/075004.
[8] R.K. Jain, S. Datta, S. Majumder, Bio-mimetic behavior of IPMC using EMG signal for a micro robot, Mech. Based Des. Struct. Mach.: Int. J. 42 (3) (2014) 398—417.
[9] R.K. Jain, S. Datta, S. Majumder, Design and control of an IPMC artificial muscle finger for micro gripper using EMG signal, Mechatronics 23 (3) (2013) 381—394.
[10] M. Shahinpoor, K.J. Kim, Ionic polymer—metal composites: IV. Industrial and medical applications, Smart Mater. Struct. 14 (1) (2004) 197—214. Available from: 10.1088/0964-1726/14/1/020.
[11] M. Billah, Z.M. Yusof, K. Kadir, A.M. MohdAli, Force measurement of non-linear IPMC using hydrodynamics analysis, in: IEEE 5th International Conference on Smart Instrumentation, Measurement and Application (ICSIMA), Songkla, Thailand, November 28—30, 2018, Available from: 10.1109/ICSIMA.2018.8688762.

[12] M. Shahinpoor, Ionic Polymer Metal Composites (IPMCs): Smart Multi-Functional Materials and Artificial Muscles, vol. 2, 2015, Royal Society of Chemistry, https://doi.org/10.1039/9781782627234.

[13] Inamuddin, S. Hussain, R.K. Jain, M. Naushad, Poly (3,4-ethylenedioxythiophene): polystyrene sulfonate zirconium (IV) phosphate composite ionomeric membrane for artificial muscle application, RSC Adv. 5 (103) (2015) 84526–84534. Available from: 10.1039/C5RA12731A.

[14] K.J. Kim, M. Shahinpoor, Ionic polymer–metal composites: II. Manufacturing techniques, Smart Mater. Struct. 12 (2003) 65–79.

[15] K. Jung, J. Nam and H. Choi Investigations on actuation characteristics of IPMC artificial muscle actuator. Sens. Actuators A Phys. 107(2003)183–192.

[16] Inamuddin, A. Khan, R.K. Jain, M. Naushad, Development of sulfonated poly (vinyl alcohol)/polpyrrole based ionic polymer metal composite (IPMC) actuator and it's characterizations, Smart Mater. Struct. 24 (9) (2015) 095003. Available from: 10.1088/0964-1726/24/9/095003.

[17] R.Y.M. Huang, P. Shao, C.M. Burns, X. Feng, Sulfonation of poly(ether ether ketone) (PEEK): kinetic study and characterization, J. Appl. Polym. Sci. 82 (2001) 2651–2660.

[18] V.K. Nguyen, Y.A. Yoo, Novel design and fabrication of multilayered ionic polymer-metal composite actuators based on Nafion/layered silicate and Nafion/silica nanocomposites, Sens. Actuators B Chem. 123 (2007) 183–190.

[19] M. Shahinpoor, Y. Bar-Cohen, T. Xue, J.O. Simpson, J. Smith, Ionic Polymer-Metal Composites (IPMC) as Biomimetic Sensors and Actuators-Artificial Muscles, ACS Chapter XX ACS Book series, 1998, pp. 1–17.

[20] Y. Bar-Cohen, Artificial muscles using electroactive polymers (EAP): capabilities, challenges and potential, electroactive polymer (EAP) actuators as artificial muscles: reality, potential, and challenges, 2004, https://doi.org/10.1117/3.547465.

[21] Y. Bar-Cohen, Electroactive polymers as artificial muscles, Chapter 4, Compliant Structures in Nature and Engineering, vol. 20, WIT Transactions on State of the Art in Science and Engineering, 2005, pp. 69–81.

[22] M. Yamakita, N. Kamamichi, Y. Kaneda, K. Asaka, Z.W. Luo, IPMC linear actuator with redoping capability and its application to biped walking robot, in: IFAC Mechatronic Systems, Sydney, 2004, pp. 353–358.

[23] J. Jung, B. Kim, Y. Tak, J.O. Park, Undulatory Tadpole Robot (TadRob) using ionic polymer metal composite (IPMC) actuator, in: IEE/RSJ International Conference on Intelligent Robots and Systems, LasVegas, Nevada, October 27–31, 2003, pp. 2133–2138.

[24] B. Kim, D.H. Kim, J. Jung, J.O. Park, A biomimetic undulatory tadpole robot using ionic polymer–metal composite actuators, Smart Mater. Struct. 14 (2005) 1–7. Available from: 10.1088/0964-1726/14/0/000.

[25] B. Tondu, Artificial Muscles for Humanoid Robots, Humanoid Robots: Human-like Machines, Edited by: Matthias Hackel, ITech, Vienna, Austria, 2007, pp. 90–123, ISBN 978-3-902613-07-3.

[26] N. Kamamichi, M. Yamakita, T. Kozuki, K. Asaka, Z.W. Luo, Doping effects on robotic systems with ionic polymer metal composite actuators, Adv. Robot. 21 (1–2) (2007) 65–85. Available from: 10.1163/156855307779293634.

[27] K.J. Kim, S. Tadokoro, Electroactive Polymers for Robotic Applications: Artificial Muscles and Sensors, Springer-Verlag London Limited, 2007.

[28] Z. Chen, X. Tan, A control-oriented and physics-based model for ionic polymer–metal composite actuators, IEEE/ASME Trans. Mechatron. 13 (5) (2008) 519–529.

[29] Z. Chen, S. Shatara, X. Tan, Modeling of biomimetic robotic fish propelled by an ionic polymer–metal composite caudal fin, IEEE/ASME Trans. Mechatron. 15 (3) (2010) 448–459.

[30] B. Koo, D.S. Na, S. Lee, Control of IPMC Actuator using self-sensing method, IFAC Proc. 42 (3) (2009) 267–270.

[31] J.S. Najem, Design and Development of a Bio-inspired Robotic Jellyfish That Features Ionic Polymer Metal Composites Actuators (M.S. thesis), Virginia Polythechnic Institute and State University, United States, April 2012.

[32] W. Kim, H.J. Kim, Nonlinear learning control of ionic polymer metal composites, in: 11th IFAC International Workshop on Adaptation and Learning in Control and Signal Processing, Caen, France, July 3–5, 2013, pp. 233–238.

[33] R. Mutlu, G. Alici, W. Li, Electroactive polymers as soft robotic actuators: electromechanical modeling and identification, in: IEEE/ASME International Conference on Advanced Intelligent Mechatronics (AIM), Wollongong, Australia, July 9–12, 2013, pp. 1096–1101.

[34] X. Chen, Adaptive Control for Ionic Polymer-Metal Composite Actuator Based on Continuous-Time Approach, in: 19th World Congress the International Federation of Automatic Control, Cape Town, South Africa, August 24–29, 2014, pp. 5073–5078.

[35] A.B. Sun, D. Bajon, J.M. Moschetta, E. Benard, C. Thipyopas, Integrated static and dynamic modeling of an ionic polymer–metal composite actuator, J. Intell. Mater. Syst. Struct. (2014). Available from: 10.1177/1045389X14538528.

[36] Y. Wang, T. Sugin, Ionic polymer actuators: principle, Fabrication and Applications, InTech, 2016, pp. 39–56. Available from: http://doi.org/10.5772/intechopen.75085.

[37] S.D. Peterson, M. Porfiri, Energy Exchange Between Coherent Fluid Structures and Ionic Polymer Metal Composites, Toward Flow Sensing and Energy Harvesting, RSC Smart Materials No. 18 Ionic Polymer Metal Composites (IPMCs): Smart Multi-Functional Materials and Artificial Muscles: Edited by Mohsen Shahinpoor, 2 (2016) 1–18.

[38] K. Kaneto, Research trends of soft actuators based on electroactive polymers and conducting polymers, J. Phys. Conf. Ser. 704 (2016) 012004 (1–9), doi:10.1088/1742-6596/704/1/012004.

[39] Y. Yan, Novel Electroactive Soft Actuators Based on Ionic Gel/Gold Nanocomposites Produced by Supersonic Cluster Beam Implantation, University of Milan, Milan, 2016.

[40] J.D. Carrico, T. Tyler, K.K. Leang, A comprehensive review of select smart polymeric and gel actuators for soft mechatronics and robotics applications: fundamentals, freeform fabrication, and motion control, Int. J. Smart Nano Mater. 8 (4) (2017) 144–213. Available from: 10.1080/19475411.2018.1438534.

[41] J.D. Carrico, N.W. Traeden, M. Aureli, K.K. Leang, Fused filament 3D printing of ionic polymer-metal composites (IPMCs), Smart Mater. Struct. 24 (2015) 125021.

[42] W. Hong, A. Almomani, Y. Chen, R. Jamshidi, R. Montazami, Soft ionic electroactive polymer actuators with tunable non-linear angular deformation, Materials 10 (664) (2017) 1–15. Available from: 10.3390/ma10060664.

[43] Z. Chen, A review on robotic fish enabled by ionic polymer–metal composite artificial muscles, Robot. Biomim. 4 (24) (2017) 1–13. Available from: https://doi.org/10.1186/s40638-017-0081-3.

[44] G.D. Pasquale, S. Graziani, C. Gugliuzzo, A. Pollicino, Ionic polymer-metal composites (IPMCs) and ionic polymer-polymer composites (IP2Cs): effects of electrode on mechanical, thermal and electromechanical behaviour, AIMS Mater. Sci. 4 (5) (2017) 1062–1077. Available from: 10.3934/matersci.2017.5.1062.

[45] Qi Shen, Theoretical and Experimental Investigation on the Multiple Shape Memory Ionic Polymer-Metal Composite Actuator (Ph.D. thesis), Mechanical Engineering, University of Nevada, Las Vegas, NV, October 2017.

[46] A. Khan, R.K. Jain, B. Ghosh, Inamuddin, A.M. Asiri, Novel ionic polymer–metal composite actuator based on sulfonated poly (1,4-phenyleneetherether-sulfone) and poly vinylidenefluoride/ sulfonated graphene oxide, RSC Adv. 8 (2018) 25423–25435.

[47] A. Ansaf, T.H. Duong, N.I. Jaksic, J.L. DePalma, A.H. Al-Allaq, B.M. Deherrera, et al., Influence of humidity and actuation time on electromechanical characteristics of ionic polymer-metal composite actuators, in: 28th International Conference on Flexible Automation and Intelligent Manufacturing (FAIM2018), Columbus, OH, June 11–14, 2018, pp. 960–967.

[48] J. Liao, Soft-Matter Artificial Muscle by Electrochemical Surface Oxidation of Liquid Metal (M.S. thesis), Carnegie Mellon University, United States, August 2018.

[49] M. Park, J. Kim, H. Song, S. Kim, M. Jeon, Fast and stable ionic electroactive polymer actuators with PEDOT:PSS/(Graphene–Ag-Nano wires) nano composite electrodes, Sensors 18 (2018). Available from: 10.3390/s18093126. 3126 (1–14).

[50] T.N. Nguyen, K. Rohtlaid, C. Plesse, G.T.M. Nguyen, C. Soyer, et al., Ultrathin electro chemically driven conducting polymer actuators: fabrication and electro chemomechanical characterization, Electrochim. Acta 265 (2018) 670–680. Available from: 10.1016/j.electacta.2018.02.003.

[51] M. Annabestani, M. Fardmanesh, Ionic electro active polymer-based soft actuators and their applications in microfluidic micropumps, microvalves, and micromixers: a review, arXiv preprint arXiv:1904.07149, 2019 (2019) 1–32.

[52] J. Zhang, J. Sheng, C.T. O'Neill, C.J. Walsh, R.J. Wood, J.H. Ryu, et al., Robotic artificial muscles: current progress and future perspectives, IEEE Trans. Robot. 35 (3) (2019) 761–781.

[53] C. Zhang, B. He, Z. Wang, Y. Zhou, A. Ming, Application and analysis of an ionic liquid gel in a soft robot, 2019 (2019) 14, Article ID 2857282, https://doi.org/10.1155/2019/2857282.

[54] C. Zhang, B. He, A. Ding, S. Xu, Z. Wang, Y. Zhou, Motion simulation of ionic liquid gel soft actuators based on CPG control, Comput. Intell. Neurosci. 2019 (2019), 11, Article ID 8256723, https://doi.org/10.1155/2019/8256723.

[55] B. Andò, S. Graziani, M.G. Xibilia, Low-order nonlinear finite-impulse response soft sensors for ionic electroactive actuators based on deep learning, IEEE Trans. Instrum. And. Meas. 68 (5) (2019) 1637–1646.

[56] D. Kongahage, J. Foroughi, Actuator materials: review on recent advances and future outlook for smart textiles, Fibers 7 (3) (2019) 1–23. fib7030021-1-fib7030021-24.

[57] V.H. Nguyen, et al., Electroactive artificial muscles based on functionally antagonistic core–shell polymer electrolyte derived from PS-b-PSS block copolymer, Adv. Sci. 6 (2019). 1801196 (1–7).

[58] N. Sharma, S. Mukherjee, Fabrication, characterization and application of ionic polymer metal composite, COLLOQUIUM, in: Conference on Mechanical Engineering and Technology, Indian Institute of Technology (BHU), Varanasi, April 06–07, 2019, pp. 37–41.

[59] Z. Chen, P. Hou, Z. Ye, Robotic fish propelled by a servo motor and ionic polymer-metal composite hybrid tail, J. Dyn. Syst. Meas. Control. 141 (2019). 071001(1–11).

[60] E.S. Tabatabaie, M. Shahinpoor, Novel configurations of slit tubular soft robotic actuators and sensors made with ionic polymer metal composites (IPMCs), Robot. Autom. Eng. J. 3 (4) (2018). Available from: 10.19080/RAEJ.2018.03.555616. 555616 (1–10).

[61] E.S. Tabatabaie, M. Shahinpoor, Artificial soft robotic elephant trunk made with ionic polymer-metal nanocomposites (IPMCs), Int. Robot. Autom. J. 5 (4) (2019) 138–142.

[62] W.H. MohdIsa, A. Hunt, S.H. HosseinNia, Sensing and self-sensing actuation methods for ionic polymer–metalcomposite (IPMC): a review, Sensors 19 (2019). Available from: 10.3390/s19183967. 3967 (1–36).

[63] W.H. MohdIsa, A. Hunt, S.H. HosseinNia, Active sensing methods of ionic polymer metal composite (IPMC): comparative study in frequency domain, in: 2019 IEEE International Conference on Soft Robotics (RoboSoft 2019), Seoul, April 14—18, 2019, pp. 546—551.

[64] A. Khan, I. Inamuddin, R.K. Jain, Easy, operable ionic polymer metal composite actuator based on a platinum-coated sulfonated poly(vinyl alcohol)-polyaniline composite membrane, J. Appl. Polym. Sci. 133 (33) (2016) (43787)(1—9).

[65] V. Panwar, S.Y. Ko, J.O. Park, S. Park, Enhanced and fast actuation of fullerenol/PVDF/PVP/PSSA based ionic polymer metal composite actuators, Sens. Actuators B Chem. 183 (2013) 504—517.

[66] A. Inamuddin, M. Khan, Luqman, A. Dutta, Kraton based ionic polymer metal composite (IPMC) actuator, Sens. Actuators A Phys. 216 (2014) 295—300.

[67] S. Nemat-Nasser, Micromechanics of actuation of ionic polymer-metal composites, J. Appl. Phys. 92 (2002) 2899—2915.

[68] A. Inamuddin, R.K. Khan, Jain, M. Naushad, Study and preparation of highly water-stable polyacrylonitrile-kraton-graphene composite membrane for bending actuator toward robotic application, J. Intell. Mater. Syst. Struct. 27 (2016) 1534—1546.

[69] Q. Zhang, Y. Li, Y. Feng, W. Feng, Electropolymerization of graphene oxide/polyaniline composite for high-performance supercapacitor, Electrochim. Acta 90 (2013) 95—100.

[70] T.A. Zawodzinski, Water uptake by and transport through nafion® 117 membranes, J. Electrochem. Soc. 140 (1993) 1041.

[71] M. Rajagopalan, J.H. Jeon, I.K. Oh, Electric-stimuli-responsive bending actuator based on sulfonated polyetherimide, Sens. Actuators B Chem. 151 (2010) 198—204.

[72] K.R. Prasad, N. Munichandraiah, Fabrication and evaluation of 450F electrochemical redox supercapacitors using inexpensive and high-performance, polyaniline coated, stainless-steel electrodes, J. Power Sources 112 (2002) 443—451.

[73] Inamuddin, A. Khan, R.K. Jain, A.M. Asiri, Thorium (IV) phosphate-polyaniline composite based hydrophilic membrane for bending actuator application, Polym. Eng. Sci. 57 (3) (2017) 258—267.

CHAPTER 3

Soft actuators and their potential applications in rehabilitative devices

Alexandrea Washington, Justin Neubauer and Kwang J. Kim
Department of Mechanical Engineering, University of Nevada-Las Vegas, Las Vegas, Nevada, NV, United States

3.1 Introduction

Rehabilitation is the restoration, especially by therapeutic means, to an improved condition of physical function [1]. This includes the restoration of something damaged or deteriorated due to a prior condition. Living with a physical ailment is not only debilitating physically, it can also be detrimental to mental health and cause increased stress. Limited mobility can be decrease a person's quality of life. The highest quality of life may be achieved when a person is able to adapt and grow into a new state of life and reclaim their health. The ability to adapt and adjust to a new health condition will determine the eventual quality of life for the person, and the amount of empowerment he or she will have over daily life [2]. Also, there is an increased cost and decreased quality of life if a person is not able to physically perform daily tasks as they normally would [2]. Physical impairments or deterioration can occur through a number of means such as accidents, activity related injuries, disease, and old age. However, rehabilitation is a very common diagnosis for physical impairments. For example, strokes in older adults are a main cause of adult disability, however, by partaking in training or rehabilitation, in which the body is repeatedly trying to accomplish some task, recovery of bodily function is best achieved [3–6]. Research in neurological and nervous system disorders show that rehabilitation and task-oriented training is the most effective method of managing the disease right after medical treatment [7], which can be costly and painful. Even people of advanced age benefit from physical rehabilitation as some body parts begin to deteriorate, it helps individuals develop, maintain, or restore movement [8].

Because physical impairments are so common, rehabilitation is one of the best methods to enhance and restore functional ability and quality of life.

Conventional rehabilitation is incredibly helpful, but it too has some drawbacks. Commonly, rehabilitative techniques need to be continued after the allotted time; however, continued therapy can be expensive [9]. There is a need for improvement in traditional rehabilitation; that advancement can be found in robotics. Robotic assistance in rehabilitation can curb some of the problems of current rehabilitation. With robotic assistance, the rehabilitation continues after leaving the doctor, thus increasing the impact the therapy is making on the patient [7,10]. Robotic assistance can also be made portable, for use in the doctor's office or at home [10]. It can also be more cost-effective because it is a one-time purchase; a better alternative to continued doctor's visits over months or even years [7,10]. With constant rehabilitation, the patient will have an improved experience, improved treatment, and lasting effects from a better therapy process.

One common aid in rehabilitation is the use of an exoskeletal brace. This allows the user to exercise the impaired area of their body while still being supported. It is important that the user should not be harmed in the process of rehabilitation. Exoskeletal brace systems and other traditional robotic systems do improve the rehabilitation process but can often be confining and bulky [11–14]. Thus one important feature of these mechanisms is compliance, and the emerging field of soft robotics solves this problem. Soft robotics is a field in which robotic devices and structures use compliant materials to better equip robotic systems for human and animal use [13]. Soft robotic structures often use materials that work better with the human body, whether they conform to the body or can be placed into the body. Humans mostly comprise soft tissues which help our system function, thus if robotic assistance is necessary, the robotic system must be able to integrate into the environment of the human body [13]. Also, an innate feature of soft robotic structures is that the systems can be reduced in size but give good output performance, such as large deformations and blocking forces [14].

It is essential to optimize the rehabilitation process in order to maximize the rehabilitation people may receive. This can be achieved through soft robotics. As mentioned previously, soft robotics is defined as so due to the method of actuation. Actuation systems can come in many sizes and forms and this will ultimately determine how it is used in a larger system. The driving force of the systems, actuators, can be broadly categorized into electroactive polymer

(EAPs) actuators and hybrid actuators (HAs), both will be further defined within this text. Within the category of EAPs exists ionic electroactive polymers (Ionic EAPs), which includes electrically activated hydrogels and ionic polymer—metal composites (IPMCs) [15,16], and polyvinyl chloride (PVC) gels [17]. Under the HAs types exist the hydraulically amplified self-healing electrostatic (HASEL) actuator, which is the combination of dielectric soft robotic principles and hydraulic actuation [18,19], the pneumatic and hydraulic-based actuators that use a compliant shell and pneumatic or hydraulic principles for actuation [20,21], the braided pneumatic muscle [22], and nylon coils which use the principles of thermal—mechanical actuation [23,24]. Improvement of the existing therapeutic mechanisms will provide maximum restoration of functional abilities. This chapter will attempt to investigate soft robotics as a means to improve existing rehabilitative mechanisms, in order to optimize the rehabilitative process. An overview of basic soft actuation mechanisms will be covered and a few case studies will be provided to demonstrate the effectiveness of compliant soft mechanisms in rehabilitative applications.

3.2 Overview of soft robotic actuators

Actuators are needed in order to drive or aid in the motion of the rehabilitative mechanism. Traditional robotics involves rigid components and is not ideal for human/robot interactions. These rigid components do not naturally follow the form of the human body and may cause further damage to patients in the rehabilitation process. Soft actuation mechanisms involve soft materials and components that are more suitable for human/robot interactions. These soft materials more easily conform to human motion and allow for deviations during the motion of the mechanism. This natural compliance in soft materials prevents injury when interacting with the human body. Soft robotics has been used to mimic biological systems, which is known as biomimetics or biomimicry [25]. This may include a small portion of a biological system, such as cilia, or larger biological systems [26]. Soft robotics has also been used for artificial muscles [27]. The following actuators are a small sample within the soft robotics field that may be suitable for rehabilitation purposes.

3.2.1 Electroactive polymers

Electroactive polymers (EAPs) include a wide range of materials that exhibit a change in shape when stimulated by an electric field [28]. EAPs also include soft sensing mechanisms, but soft actuation types will be outlined in this chapter. These EAPs exhibit a range of deformation types and deformation magnitudes over a range of electrical inputs. EAPs include piezoelectrics, shape memory alloys, magnetostrictive materials, along with other materials. EAPs are sought after soft robotic mechanisms due to their large deformations and high response time, along with their lightweight and inexpensive production. EAPs can be split into two main categories based on the working principle: ionic and electronic EAPs. Ionic EAPs use the displacement of ions as the mode of actuation. These actuators generally have very low driving voltages with low response times. Electronic EAPs rely on high electric fields to change the electromechanical structure, and thus change the shape of the material. These EAPs also tend to have a faster response time.

3.2.1.1 Polyvinyl chloride gels

Polyvinyl chloride (PVC) gels are electroactive non-ionic gels that use a creep deformation as the method of actuation [29]. These gels are fabricated by plasticizing the PVC with dibutyl adipate (DBA). To amalgamate the PVC with the DBA, tetrahydrofuran is added to the solution as a solvent [17]. PVC gels use a creeping mechanism of deformation under an electric field. The gel is placed between two electrodes. When the electrodes are activated, electrons can penetrate the gel from the cathode and migrate toward the anode. As the charge builds on the anode, the electrostatic adhesiveness gradually forces the gel toward the anode surface [17]. Different anode geometries are being investigated to determine maximum deformation between cathode and anode surfaces [25]. For example, when using a steel mesh electrode, the gel can use this creep deformation to deform around each steel cylindrical section to penetrate into the anode, potentially creating higher deformation compared to a flat steel surface [25] (Fig. 3.1).

Based on the composition of PVC gel, the force output can be as great as 3 kPa [17,29]. PVC gels are generally fabricated with a thickness of 1 mm or less [17,29]. A single layer consisting of a cathode, PVC gel, and anode, typically actuates less than 1 mm but is dependent on doping agents within the PVC gel [25]. Due to the small nature of PVC gels, they are often stacked into many layers with ordering similar to cathode, PVC gel, anode, PVC gel, cathode, etc. [25] The power requirements for this actuator include high voltages, generally in the kilovolt range, and low current, generally in the microamp range [25]. In relation to

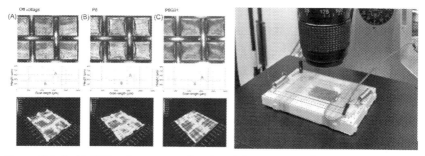

Figure 3.1 (Left) PVC gel actuators showing deformation under an applied electric field, (A) without an applied electric field. (B) with an applied electric field, (C) with an applied electric field (1.5 kV with DC) and (right) an experimental setup.

biocompatibility, the plasticizer is toxic, so it cannot be used without some protective layer between the actuator and the human body [29]. Because of these qualities of PVC gels, these soft robotic mechanisms are tailored for use outside of the body. The small, compact, and scalable size of the PVC gel mechanisms make them ideal for use in portable systems. The electrical input required for these PVC gels is generally high, so the power requirements do create a need for a more advanced electrical system or increase performance of these gels in order to create widespread use. Due to the size, response time, and force output, PVC gel systems are a proper candidate for use in rehabilitation.

3.2.2 Ionic electroactive polymers
3.2.2.1 Ionic hydrogels

Ionic hydrogels are a class of hydrogels that are hydrated with an electrolytic solution in order to produce actuation [30]. The ionization process includes an electrolyte solution such as lithium chloride or sodium chloride. This salt solution is mixed in the gel solution and is then cured to allow for a deposition-type process within the gel [31]. The principle of this actuator is the migration of ions due to the induced electric field. In the electrolyte solution, the charged salt ions become hydrated when introduced to water. The gel is then excited by an electrical source creating an electric field. The hydrated ions, including both anions and cations, migrate to one side of the gel. This causes a contraction in the lateral direction of the hydrogel and an elongation in the longitudinal direction [30].

Hydrogels are generally fabricated with a thickness in the millimeter to centimeter range [30]. The voltage range of actuation is generally from

6 kV up to 18 kV [30]. Given this voltage input a hydrogel can generally reach greater than 100% strain, in fact an area strain of about 170% can be obtained. There is also an exponential increase in strain as the voltage increases [30]. Hydrogels have excellent displacement capabilities and response times; however, it is important to note that as the frequency input increases, the response of the actuator can decrease [30]. Hydrogels have the ability to conform to living tissues and cells on the molecular level [30]. Given the high biocompatibility and ability to conform to living cells, these hydrogels can be used outside or inside the body. However, the hydration can significantly affect the results of actuation [31]. This class of hydrogels are a good candidate for use in rehabilitation due to their biocompatibility but do have a large driving voltage which may affect the overall size of the system, as the electrical system may become cumbersome.

3.2.2.2 Ionic polymer–metal composite

Ionic polymer-metal composits (IPMCs) are Ionic EAPs that are composed of a perfluorinated polymer membrane such as Nafion, Flemion, or Aquivion with chemically plated electrodes [15,16,32–36]. The perfluorinated membrane is a porous polymeric material that allows for ion transport. The electrodes are typically made from noble metals such as platinum or gold due to their chemical stability [15,16,32,34,36,37]. Using noble metals does provide some drawbacks, such as cracking of the electrodes and degradation of electrodes over time, resulting in decreased performance, and additional processing costs [32,38]. Thus there have been many variations made on IPMCs to enhance certain features such as displacement, blocking force, and biocompatibility. The variations of IPMCs include the use of ultrathick IPMCs, carbon-based or doped electrodes, electrodeless IPMCs, IPMCs that utilize shape memory actuation principles, and IMPCs in which the geometry has been altered. In the case of the ultrathick IPMCs, the perfluorinated membranes are stacked, such that the thickness is five times larger than a standard IPMC [39]. With this IPMC variation, incredibly large forces are generated; these actuators are able to displace about 125 g [39] (Fig. 3.2).

Another beneficial variation is to use carbon electrodes, doped graphene electrodes with sulfur and oxygen (allowing for n-type doping, such that the electrodes have excess electrons [38]), or doping the membrane with electroactive nanoparticles [40]. This would cause more electrons to migrate from the anode causing a greater electric field, and in

Figure 3.2 An IPMC in action (left) off state and (right) in action.

turn larger actuation. In the case of doping the membrane itself, there are improved electrical and mechanical properties of the actuator caused by the electrostatic effects of the dispersed particles [40]. By using the carbon-based or doped graphene electrodes, the actuator can be used in air and is not just limited to water like a typical IPMC [38].

Some research has gone into electrodeless IPMCs using perfluorinated membranes without depositing noble metal electrodes. Actuation performance relies solely on the ion transport within the membrane [32]. This is particularly of importance when considering how the electrodes can drastically affect the performance. In the case of typical IPMCs, the electrodes dictate the performance, in that their structure determines the absorption of the ions, also their flexibility, conductivity, and surface morphology play a crucial role [26]. Electrodeless IPMCs also decrease the cost of fabrication, the electrode implantation process can be laborious and use multiple materials due to the fabrication complexities; however, with no electrodes there is only a single material with a simple fabrication method [26]. However, these actuators are not actually IPMCs due to the absence of the metal—composite layer, but they still consider the principle methods of IPMC actuation and are, in a sense, an extension of the IPMC field.

Another interesting field in IPMC research is shape memory. IPMCs have shape memory features, meaning that the material can memorize a permanent shape, be deformed into another shape, and then return to the original shape when exposed to external stimulus (temperature, electricity, etc.) [33]. The materials used for IPMCs are inherently soft and resilient, making it suitable for shape memory application when heat cycles are applied. Nafion has multiple shape memory modes and is able to achieve a series of complex shapes [33,34]. This potentially has applications in rehabilitation but relies on temperature change as the mode of actuation for the material. This limits its use almost exclusively to external use as the body has a regulated temperature.

Cylindrical IPMCs are also of interest to the field of rehabilitation. Cylindrical IPMCs use the same principles as a standard rectangular IMPC, the only difference is the geometry. However, this change can affect the flexibility, method of fabrication, and displacement output [41,42]. Because of the circular cross-section, the actuator can retain more flexural rigidity, thus leading to small displacements. Also, the new geometry can increase the difficulty of fabrication [41,42]. Considering these drawbacks, this novel research can still be beneficial to the field of rehabilitation.

As mentioned previously, the fabrication process for manufacturing IPMCs is incredibly labor-intensive and requires extensive lab work [26,37,42,43]. IPMC fabrication can be broken down into four categories (in some cases five categories): membrane fabrication, surface preparation, activation, the surface electroding, and ion implantation. It is common to purchase perfluorinated membranes such as Nafion and Aquivion; however, membranes can be fabricated by melting down pellets and molding them to the desired shape and size. Once a membrane is acquired, the surface must be prepared for the activation and plating process. During the surface preparation stage, the surface is leveled and roughened for the optimal penetration of the noble metal. The membrane is then activated so the ion migration can occur. However, it is important to note that many membranes can be bought preactivated. The membrane is then plated through a chemical reduction process [37,43]. Finally, the IPMC is placed in an electrolyte bath so that the ions of the salt will be implanted and act as the hydrated cations. In general, these actuators rely on migration of hydrated cations under an induced electric field. This migration causes a swelling in one side of the actuator, resulting in a bending motion. The ultrathick IPMCs, the shape memory IPMCs, and cylindrical IPMCs are generally made using the same or a similar methodology [32,33,39,41,42].

Generally, IPMCs have low activation voltages, as low as two volts, and have a blocking force of about 1.55 g [43]. One must consider that the blocking force is affected by the length of the IPMC; the blocking force will decrease as the length increases. IPMCs have some biocompatibility and may have applications inside or outside of the human body. The hydration of the IPMC can greatly affect the performance and is used almost exclusively in aqueous environments. For this reason, IPMCs are very suitable for applications inside of the body. IPMCs are well suited for small system designs and have a very low driving voltage, usually less than 5 V [39,43,44].

3.2.3 Hybrid actuators

Hybrid soft actuators apply classical mechanics such as thermal expansion, fluid mechanics, and electrostatics on soft materials such as rubbers, fabrics, and other polymers. These actuators use classical robotics techniques and apply them in an innovative way to soft materials. Some examples of these hybrid soft actuators include pneumatics, hydraulics, and nylon coil actuators.

3.2.3.1 Hydraulically amplified self-healing electrostatic actuator

One hybrid soft actuator is the hydraulically amplified self-healing electrostatic (HASEL) actuator, developed by the Keplinger Group of CU Boulder. These actuators use principles of electrostatics and hydraulics to cause a displacement in the actuator. Electrostatic actuation is the configuration of two parallel electrodes with a dielectric in between the electrodes, the electrodes are charged with a voltage, and the charge of the dielectric aligns with the electric field generated by the electrodes. In the case of the HASEL actuators, a liquid dielectric sits between two compliant electrodes, contained by a flexible polymer shell [18,19]. In general, the liquid dielectric has a relative permittivity of 2.0−3.1. When the electrodes are charged and critical voltage is reached, the compliant electrodes and film gradually collapse toward each other from one corner of the electrode to the other [45,46]. A hydraulic actuation system uses the pressure of a fluid to cause actuation. In the case of the HASEL actuators, as the electrodes collapse or "zip" toward each other, the fluid is displaced [47]. When the electrodes buckle at the critical voltage and close together, a majority of the fluid is pushed into the area not covered by an activated electrode. A thin layer of the liquid dielectric remains between the electrode [47], this is caused by the continued application of the electric field creating a field within the dielectric, thus preventing the electrodes from touching and short circuiting (Fig. 3.3).

The fabrication of the HASEL actuator is simple and cost-effective [18,19]. As mentioned previously, there are only three materials that are needed for the actuator. A thin polymer shell can be fabricated or purchased at a low cost and the liquid dielectric can be an oil, such as mineral oil or transformer oil [18,19,48]. The electrodes can be made from a number of flexible conductive materials, such as ionic hydrogels, aluminum foil, or be a flexible carbon-based material. The performance of the actuator is heavily reliant on the geometry of the packet. HASEL actuators require a high

Figure 3.3 (Left) Front view of the HASEL actuator as it is activated. (Right) Side view of the HASEL actuator as it is activated.

activation voltage in the range of 5–25 kV at a low current draw. At the maximum voltage, the maximum actuation strain seen is about 80% with a 250 g load [18,19]. Because of their size and geometry, these actuators can be stacked to provide an even greater output force and displacement. As mentioned previously, the HASEL actuator uses a thin film packet, which may cause issues if used inside of the body due to leakage of the dielectric fluid. But overall, the HASEL actuators produce large deformations under a large range of frequencies.

3.2.3.2 Soft pneumatic and hydraulic actuators

Soft pneumatics and hydraulics are another form of hybrid soft actuation. These actuators usually consist of a silicone-based material with chambers inside of the structure. These chambers fill with air or a liquid for pneumatic and hydraulic actuation, respectively [14,20,21]. These types of actuators rely completely on their geometry to provide actuation. There also may be strain-limiting layers within the structure of the actuator. These strain-limiting layers are often made from other materials, such as stiffer rubbers or fibers that prevent strain in a localized area or direction of the strain limiting fiber [20,21]. These actuators are commonly cast in a mold using a multitude of silicone-based materials. These actuators can range greatly in size, as well as, force output and power requirements. For a fully flexed state, the pressure in the actuator is 345 kPa, other states of motion can be achieved at lower input pressures of 275 kPa. However, these systems require pumps as well as a power system, so these systems may become cumbersome. Due to the leakage possibility within the actuator, these actuators should be limited to external use for rehabilitative applications (Fig. 3.4).

Braided pneumatic actuators are very similar to soft pneumatic actuators but are made from braided fibers rather than soft rubbers. In general,

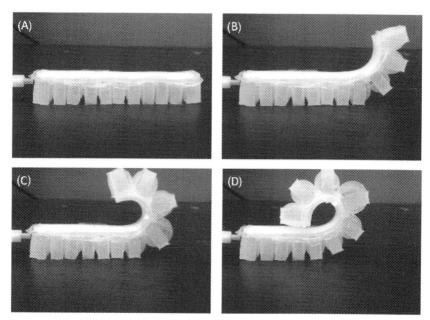

Figure 3.4 Example of bending pneumatic actuator fabricated from soft elastomeric material actuating from A–D.

these actuators produce much higher force output, but require a much larger pump. These actuators tend to produce linear actuation and are much closer to traditional robotics compared to other soft robotic actuation types. These actuators have similar issues with leakage and should be limited to external body use.

3.2.3.3 Nylon-based coiled polymer actuator

Nylon-based coils are another type of soft actuation, more specifically a type of coiled polymer actuator (CPA). CPA actuators use one or more coiled fibers to create a heat-driven soft linear actuator. These actuators use commercially produced polyethylene and nylon fibers, which provide reversible thermal contraction in the fiber direction, large thermal expansion in all directions, and anisotropic dimensional changes, which allow for enhanced muscle-like performance [24]. To be more specific, these fibers have negative coefficients of thermal expansion, thus they contract when heated; by coiling the nylon, tensile contractions are produced [24,32]. The exact mechanism of actuation is such that when the fiber is twisted the polymer chain is oriented helically. The twisted fiber then tries to simultaneously contract and expand radially when heated. Thus the fiber wants to untwist, but instead causes a change in the coil length (Fig. 3.5).

The thermal expansion of the CPA can occur by supplying heat in two ways. The first type of CPA introduces heat into the system in the traditional way. This usually includes the use of a working fluid to provide the heat to and from the CPA. The second type of CPA introduces heat through electrodes by way of Joule heating or heating a conductive

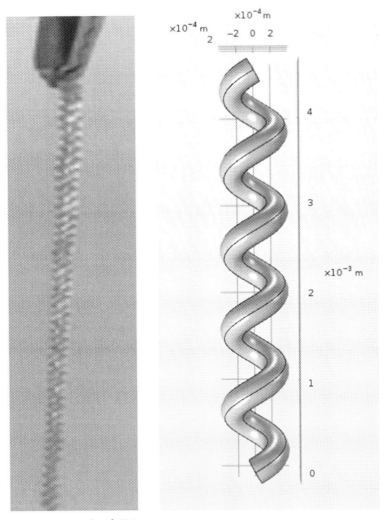

Figure 3.5 An example of CPA actuator.

material by passing an electric current through it [32]. A conductive paint can be used to coat the nylon coils and cause Joule heating, and thereby the actuation of the coiled nylon or polyethylene [32]. As mentioned previously, nylon CPAs have a huge advantage of being able to be fabricated from material that is readily available, and even can be fabricated from off-the-shelf fishing line. These actuators provide fast response times and large power densities. In fact the CPAs can exceed the maximum force output of human muscles [24]. Due to the heat impact on the human body, this type of actuation should be limited to external use. These systems may become cumbersome if heat is being introduced in the traditional way.

The soft robotic actuators discussed are just a few examples that are suitable for rehabilitation applications. The world of soft robotic actuators is fairly extensive and each of the branches of soft robotic actuators, EAPs, Ionic EAPs, and hybrid actuator systems contain a number of soft actuation types that may or may not be suitable for rehabilitative applications. However, this wide range of actuators allows for a great number of systems to be designed with rehabilitation in mind.

3.3 Applications of soft robotic actuators

The actuation types mentioned above will be explored for applications in rehabilitation. There are a range of applications, from hip assistance for walking mobility, to hand and wrist exoskeletal gloves designed for everyday use. Some rehabilitative mechanisms strictly provide support, such as an exoskeletal brace. These devices are designed so that they do not include any actuation mechanisms and solely rely on the compliance of the material used in the device to provide rehabilitative assistance. These types of devices do not require soft robotic actuators because there are no actuation mechanisms. But, the following section will primarily investigate the uses of the actuators in rehabilitative devices and how they can benefit those in need of these robotic systems.

3.3.1 Hip joint support using polyvinyl chloride gels

A PVC gel soft actuator-based wearable assistance mechanism was fabricated and tested by Li et al. [49] This device aimed to assist patients with weakened musculature for daily use. It was deemed that the assistance of an artificial muscle would have a positive effect on the walking ability of the weakened leg during activities of daily life and rehabilitation exercises, such as increased walking speed and step length, and decreasing the level of muscular activity during walking [49]. The mechanism was designed to be lightweight, unlike many of the current braces and other exoskeleton designs. The device in this case focused on the motion of the hip joint while providing support to the area. The device is made of three main components: the actuation mechanism to provide tensile force, an insole force sensor to detect gait, and a portable battery and controller to provide power and appropriate assistance during walking. The actuation is done through two separate modules of PVC gel soft actuators that are arranged along the top of the thigh using structured textile belts to transmit the forces across the body to create torque on the hip joint using the contraction—expansion motion. The two PVC gel actuation modules include stacked PVC gels with mesh anodes and foil cathodes. During the swing of the hip a voltage is applied to the PVC actuation module and causes a contraction in the gel, resulting in a tensile force to support hip flexion [49]. Once the swing phase is complete, and the foot comes into contact with the ground surface, another voltage with opposite polarity is applied causing an extension in the gel. This decreases the output force to enable easy hip extension of the limb [49].

This system was tested on a human subject. The physical results showed improvement in both walking speed and step length. An overall increase in balance was found, thus contributing to more efficient walking. The rating given by the subject showed that they felt their leg was assisted and that the equipment was lightweight [49]. Overall, the system performed as intended, and could possibly be optimized for use in daily life [49].

Work done by Kim et al. uses the transparent nature of PVC, and provides the possibility of a variable focal lens using this actuator mechanism. More specifically, PVC gel-based microlenses have been researched and designed, inspired by the ciliary muscle of the human eye [29]. If this lens were scaled up to the size of a human lens, a soft robotic lens may be able to replace or rehabilitate a defective or wounded lens. Because the design is based on a human eye, it may be able to aid in the recovery process of a human lens.

3.3.2 Hydrogels and their applications

Most hydrogels are used in biomedical applications, for example, pacemakers and drug delivery systems [50,51]. Currently, they serve more as a method of treatment than a mode of rehabilitation. However, because of their use in vivo, they can be safely used as an active blood vessel occlusion system [50,51]. All in all, the use of hydrogels could be designed for use inside the body, which can lead to in vivo rehabilitation, thus getting rid of the need for exoskeletal rehabilitation systems.

3.3.3 Ionic polymer—metal composite applications

Along with the listed advantages of IPMCs, other significant advantages relative to rehabilitation applications include the ability to operate in aqueous environments, as well as, the ease of miniaturization. IPMCs may be easily miniaturized to the submillimeter scale. These advantages allow IPMCs to be excellent candidates for catheters. Much research has been done in this area [41,42]. Catheters are needed for draining fluids and the administering of drugs, food, or other materials to assist in patient healing. The goal of an active catheter is to assist surgeons in improving the safety of intravascular procedures by providing more precise manipulation and a reliable approach with force feedback compared to manual techniques [41,42]. An active catheter allows the user to actively move the tip of the catheter in order to easily traverse vessels within the body. The use of a small cylindrical type IPMC would potentially allow for multidirectional control which may assist medical professionals in their work and allow patients to recover faster from surgical procedures.

IPMCs have also been used to study artificial muscles. One such study includes an electrodeless array of IPMCs that actuates wirelessly by applying a low external electric field. The array contains a single perfluorinated EAP material. This system was designed to mimic the biological systems of cilia and flagella [26]. The unique structure of IPMCs allow for both actuation and sensing capabilities in underwater applications. Though there is no current direct use of this system in rehabilitation, there is some possibility of future use. These cilia exhibit a fast dynamic response, have high spatial resolution, and low electronics and mechanism requirements that make this suitable for some future microrehabilitation application. The fact that this array is electrodeless makes the array more biocompatible. A study found that Nafion has excellent bacterial adhesion resistance, cell adhesion, and biocompatibility. Cell adhesion is important because this can potentially allow for human cells and human tissue to grow on and around the device so that it becomes integrated with the human system. Nafion shows excellent surface solidity in cell–structure media as well as in buffer solution [52].

Overall, Nafion is suitable for biomedical applications that require biocompatibility with a reduced possibility of postoperative infections [52]. The wireless and low driving voltages allow for safe operation within the human body. The fact that this array operates in aqueous environments and is highly biocompatible makes this array an ideal material for use within the human body. It could be embedded within the body without causing infection and become integrated with the body and work as a support or material for an artificial organ or prosthetic implant. The sensing capabilities of this type of IPMC could also allow for tracking of rehabilitation progress or tracking if something is not healing correctly during the rehabilitative process.

3.3.4 HASEL actuators and artificial muscle applications

HASEL actuators have been studied for use as artificial muscles. Devices using these highly compliant actuators include grippers and artificial bicep muscles have been designed [18,19,47]. However, these actuators use a high voltage range for actuation, as such they are limited to external use. The equipment used to power these actuators is also cumbersome. The actuator itself utilizes inexpensive and readily available materials such as polymer films, transformer oil, or mineral oil as the dielectric liquid. These actuators have the ability to be used as an exoskeletal support for hand and arm rehabilitation. The low cost of the materials would allow for a low-cost rehabilitative device. However, the high power requirements could limit the patient's mobility. This is a relatively new type of actuation and has the ability to improve performance by utilizing other materials such as different films, dielectric liquids, or gases, as well as, altering geometry. There is also the potential to lower the power consumption and further improve the response time.

3.3.5 Soft pneumatic and hydraulic actuators used in rehabilitative devices for hands and arms

A soft robotic glove was created for combined assistance for at-home rehabilitation. The project aimed to create a soft robotic glove made from soft hydraulic actuators to provide improved hand function in repetitive task practice rehabilitation. This repetitive task practice in hand rehabilitation is used to increase hand strength, accuracy, and range of motion. The actuators were composed of composite elastomeric tubes with anisotropic fibers embedded within the elastomer. The method of fabrication is labor-intensive, costly, and slow, which often lead to challenges with patient compliance [21]. The fibers' purpose is twofold. The fibers both allow for reinforcement of the elastomeric cylinder as well as allow for mechanical programming of the actuator to perform the wide range of motion under fluid pressurization [21]. The application of the continuous fibers is crucial during fabrication and will dictate the motion of the composite elastomeric tube. Some deformation types that are attainable depending on fiber direction and orientation are bending, twisting, radial, and linear motions. The deformation of the elastomeric tube is limited by the strain-limiting layer, which is planar in this case, allowing for a bending motion. Water was used as the hydraulic fluid in this case because of the ease of use and it is commonly available. The system, including the hydraulically powered rehabilitative glove, the controller, pump, and fluid reserve, is worn on a belt around the waist [21]. These actuators are very similar to the PneuNets actuators [11,20]. These PneuNets are composed of channels that are filled with a fluid. Based on the actuator and channel geometry, along with the strain-limiting layers, the actuator will be able to deform.

A braided pneumatic actuator-based glove was designed for hand rehabilitation as well. This design uses McKibben actuators which consist of a rubber tube surrounded by a braided shell of fibers, which provides compliant soft linear actuators. Air is the preferred fluid for these actuators due to its low viscosity, compressibility, ease of storage, low weight, environmentally benign nature, and that it can be actuated very rapidly [21]. The strain-limiting layers in the glove were situated such that these layers continued to rest along the hand and fingers throughout the actuation cycle. This allows assistance in closing of the hand to create a fist or grip an object. This soft robotic device is nonintrusive and customizable. Because the glove is situated on the outside of the hand, the design is noninvasive and would not cause permanent discomfort. By using this type of actuator, the geometry of the actuator can be altered based on hand geometry and the pressure or actuation cycle can be customized for each actuator based on personal needs [21]. Another braided pneumatic system was designed for artificial muscles. The system combines both braided pneumatic actuator and metal hydrides. These metal hydrides drive the system using hydrogen compression. This braided artificial pneumatic muscle (BAPM) uses aramid fibers embedded in a rubber chloroprene membrane [22].

3.3.6 Coiled polymer actuators used in rehabilitative devices for wrists

A mechanism was designed to assist wrist motion for patients with wrist orthosis using conductive nylon actuators. The mechanism includes an exoskeletal wrist brace to aid in the rehabilitative process. The conductive nylon used in this study was silver plated nylon sewing thread. This is a somewhat common fabrication method of these CPAs [53]. The goal of this study was to design a lightweight, wearable solution to increase wrist performance during physical therapy. This device could also be used to assist in everyday tasks as well as provide support in physical therapy. In a sense, robotic assistance in physical therapy continues the therapy after leaving the doctor, thus increasing the impact the therapy is making on the patient [53]. The current design for this wrist brace includes a 3D printed brace in which the nylon coils are attached in such a way that the wrist is assisted in one degree of freedom, up and down flexion motion. The results show fast actuation and larger power densities than human muscle and skeletal systems can provide [53].

3.4 Conclusions

Rehabilitation is an excellent application for soft robotics. The compliant nature of soft robotics lends itself to a beneficial human−robot interaction. This allows rehabilitative mechanisms, such as exoskeletal mechanized braces, to conform to the human body. The soft aspect of these mechanisms allows for deviation during motion and conformity to the complex nature of human motion. However, some of these soft robotic actuation types are limited due to the volatility of the materials and power requirements. These types of soft robotic actuators also lend themselves to mechanisms used outside the body, such as exoskeletons or braces, due to the dangerous or cumbersome systems required for these mechanisms. Other actuation types, such as IPMCs are highly biocompatible and require very low driving power. These types of soft robotic actuation devices may lend themselves more to rehabilitation mechanisms that may be used inside the body, such as surgical tools or implants. Some mechanisms, such as the IPMC-based active tip catheter, are difficult to define in terms of rehabilitative mechanisms. However, this mechanism and other soft robotic surgical tools, aid in restoration to an improved condition of physical function, thus these tools are loosely included in this chapter.

All in all, this chapter lists some of the many devices that use soft actuators as a means of motion for rehabilitative devices. So many constraints must be considered while still maintaining the ultimate goal: restoring a person's quality of life. Some recurring themes with soft actuation include increased compliance in soft mechanisms which, in turn improves performance and safety when adhering to the human motion. The need for cumbersome electrical systems may be needed for some soft actuation types due to increased power requirements but this could also be potentially solved by increasing the efficiency of some actuator types through means such as improving the materials used in the actuator. There has been substantial research done in soft robotics in rehabilitative applications, but there is still much more to investigate.

Acknowledgments

Authors acknowledge partial financial support from the US National Science Foundation (NSF), Partnerships for International Research and Education (PIRE) Program, Grant No. 1545857. Any opinions, findings, and conclusions or recommendations expressed in this material are those of the author(s) and do not necessarily reflect the views of the NSF. Also, K.K. acknowledges the partial financial support from the Office of Naval Research (N00014-16-1-2356) and the Department of Energy Minority Serving Institution Partnership Program (MSIPP) managed by the Savannah River National Laboratory under SRNS contract TOA#0000403073. Additionally, special thanks go to Dr. Taeseon Hwang and Zachary Frank (PVC gels), Dr. Viljar Palmre (IPMC), and Robert Hunt (CPA actuator) who provided us their input.

References

[1] D.A. Umphred, Umphred's Neurological Rehabilitation, sixth ed., Elsevier/Mosby, St. Louis, MO, 2013. Web.
[2] S.H. Keus, B.R. Bloem, E.J. Hendriks, A.B. Bredero-Cohen, M. Munneke, Evidence-based analysis of physical therapy in Parkinson's disease with recommendations for practice and research, Mov. Disord. 22 (2007) 451–460. Available from: 10.1002/mds.21244.
[3] P. Langhorne, J. Bernhardt, G. Kwakkel, Stroke rehabilitation, The Lancet 377 (9778) (2011) 1693–1702, ISSN 0140-6736, https://doi.org/10.1016/S0140-6736(11)60325-5.
[4] Bruce H. Dobkin, Strategies for stroke rehabilitation, The Lancet Neurology 3 (9) (2004) 528–536, ISSN 1474-4422, https://doi.org/10.1016/S1474-4422(04)00851-8. (http://www.sciencedirect.com/science/article/pii/S1474442204008518)).
[5] P. Brukner, K. Khan, Clinical Sports Medicine, Brukner & Khan's Clinical Sports Medicine, 4th ed., McGraw-Hill Education, Sydney, NEW SOUTH WALES, Australia, 2012. In press.

[6] Diane L. Damiano, Activity, Activity, Activity: Rethinking Our Physical Therapy Approach to Cerebral Palsy, Physical Therapy 86 (11) (2006) 1534–1540. Available from: https://doi.org/10.2522/ptj.20050397.
[7] H.I. Krebs, L. Dipietro, S. Levy-Tzedek, S.E. Fasoli, A. Rykman-Berland, J. Zipse, et al., A paradigm shift for rehabilitation robotics, IEEE Eng. Med. Biol. Mag. 27 (4) (2008) 61–70. Available from: https://doi.org/10.1109/MEMB.2008.919498.
[8] S. Karinkanta, M. Piirtola, H. Sievänen, K. Uusi-Rasi, P. Kannus, Physical therapy approaches to reduce fall and fracture risk among older adults, Nat. Rev. Endocrinol. 6 (2010) 396–407. Available from: 10.1038/nrendo.2010.70.
[9] P. Polygerinos, Z. Wang, K.C. Galloway, R.J. Wood, C.J. Walsh, Soft robotic glove for combined assistance and at-home rehabilitation, Robot. Auton. Syst. 73 (2015) 135–143. Available from: https://doi.org/10.1016/j.robot.2014.08.014.
[10] P. Maciejasz, J. Escheriler, K. Gerlach-Hahn, A. Jansen-Troy, S. Leonhardt, A survey on robotic devices for upper limb rehabilitation, J. Neuroeng. Rehabil. 11 (1) (2014) 1–29. Available from: https://doi.org/10.1007/s00115-003-1549-7.
[11] R. Morales, F.J. Badesa, N. García-Aracil, J.M. Sabater, C. Pérez-Vidal, Pneumatic robotic systems for upper limb rehabilitation, Med. Biol. Eng. Comput. 49 (10) (2011) 1145–1156. Available from: https://doi.org/10.1007/s11517-011-0814-3.
[12] P. Polygerinos, S. Lyne, Z. Wang, L.F. Nicolini, B. Mosadegh, G.M. Whitesides, et al., Towards a soft pneumatic glove for hand rehabilitation, IEEE International Conference on Intelligent Robots and Systems, IEEE, 2013, pp. 1512–1517. Available from: https://doi.org/10.1109/IROS.2013.6696549.
[13] S. Kim, C. Laschi, B. Trimmer, Soft robotics: a bioinspired evolution in robotics, Trends Biotechnol. 31 (5) (2013) 287–294. Available from: https://doi.org/10.1016/j.tibtech.2013.03.002.
[14] D. Trivedi, C.D. Rahn, W.M. Kier, I.D. Walker, Soft robotics: biological inspiration, state of the art, and future research, Appl. Bionics Biomech. 5 (3) (2008) 99–117. Available from: https://doi.org/10.1080/11762320802557865.
[15] Y. Bar-Cohen, Electroactive polymers as artificial muscles: a review, J. Spacecraft Rockets 39 (6) (2008) 822–827. Available from: https://doi.org/10.2514/2.3902.
[16] A. O'Halloran, F. O'Malley, P. McHugh, A review on dielectric elastomer actuators, technology, applications, and challenges, J. Appl. Phys. 104 (7) (2008). Available from: https://doi.org/10.1063/1.2981642.
[17] N. Ogawa, M. Hashimoto, M. Takasaki, T. Hirai, Characteristics evaluation of PVC gel actuators, IEEE International Conference on Intelligent Robots and Systems, IEEE, 2009, pp. 2898–2903.
[18] E. Acome, S.K. Mitchell, T.G. Morrissey, M.B. Emmett, C. Benjamin, M. King, et al., Hydraulically amplified self-healing electrostatic actuators with muscle-like performance, Science 359 (6371) (2018) 61–65. Available from: https://doi.org/10.1126/science.aao6139.
[19] N. Kellaris, V.G. Venkata, G.M. Smith, S.K. Mitchell, C. Keplinger, Peano-HASEL actuators: muscle-mimetic, electrohydraulic transducers that linearly contract on activation, Sci. Robot. 3 (14) (2018) 1–11. Available from: https://doi.org/10.1126/scirobotics.aar3276.
[20] P. Polygerinos, S. Lyne, Z. Wang, L.F. Nicolini, B. Mosadegh, G.M. Whitesides, et al., Towards a soft pneumatic glove for hand rehabilitation, IEEE International Conference on Intelligent Robots and Systems, IEEE, 2013, pp. 1512–1517. Available from: https://doi.org/10.1109/IROS.2013.6696549.
[21] P. Polygerinos, Z. Wang, K.C. Galloway, R.J. Wood, C.J. Walsh, Soft robotic glove for combined assistance and at-home rehabilitation, Robot. Auton. Syst. 73 (2015) 135–143. Available from: https://doi.org/10.1016/j.robot.2014.08.014.

[22] A. Vanderhoff, K.J. Kim, Experimental study of a metal hydride driven braided artificial pneumatic muscle, Smart Mater. Struct. 18 (12) (2009). Available from: https://doi.org/10.1088/0964-1726/18/12/125014.

[23] P. Maciejasz, J. Escheriler, K. Gerlach-Hahn, A. Jansen-Troy, S. Leonhardt, A survey on robotic devices for upper limb rehabilitation, J. Neuroeng. Rehabil. 11 (1) (2014) 1−29. Available from: https://doi.org/10.1007/s00115-003-1549-7.

[24] C.S. Haines, M.D. Lima, N. Li, G.M. Spinks, J. Foroughi, J.D. Madden, et al., Artificial muscles from fishing line, Science 343 (2014) 868−872.

[25] T. Hwang, Z. Frank, J. Neubauer, et al., High-performance polyvinyl chloride gel artificial muscle actuator with graphene oxide and plasticizer, Sci. Rep. 9 (2019) 9658. Available from: 10.1038/s41598-019-46147-2.

[26] J.K. Kwang, V. Palmre, T. Stalbaum, T. Hwang, Q. Shen, S. Trabia, Promising developments in marine applications with artificial muscles: electrodeless artificial cilia microfibers, Mar. Technol. Soc. J. 50 (2016) 24−34. Available from: 10.4031/MTSJ.50.5.4.

[27] J. Rossiter, B. Yap, A. Conn, Biomimetic Chromatophores for Camouflage and Soft Active Surfaces, IOP Publishing Ltd., 2012.

[28] Yoseph Bar-Cohen: Electroactive Polymer (EAP) Actuators as Artificial Muscles: Reality, Potential, and Challenges. SPIE Press, Bellingham, WA, 2004.

[29] S.Y. Kim, M. Yeo, E.J. Shin, W.H. Park, J.S. Jang, B.U. Nam, et al., Fabrication and evaluation of variable focus and large deformation plano-convex microlens based on non-ionic poly(vinyl chloride)/dibutyl adipate gels, Smart Mater. Struct. 24 (11) (2015). Available from: https://doi.org/10.1088/0964-1726/24/11/115006.

[30] C. Keplinger, J. Sun, C.C. Foo, P. Rothemund, G.M. Whitesides, Z. Suo, Stretchable, transparent, ionic conductors, Science 341 (2013) 984−988.

[31] Y. Bai, B. Chen, F. Xiang, J. Zhou, H. Wang, Z. Suo, Transparent hydrogel with enhanced water retention capacity by introducing highly hydratable salt, Appl. Phys. Lett. 105 (2014). Available from: https://doi.org/10.1063/1.4898189.

[32] K.J. Kim, V. Palmre, T. Stalbaum, T. Hwang, Q. Shen, S. Trabia, Promising developments in marine applications with artificial muscles: electrodeless artificial cilia microfibers, Mar. Technol. Soc. J. 50 (5) (2016) 24−34. Available from: https://doi.org/10.4031/MTSJ.50.5.4.

[33] Q. Shen, V. Palmre, K.J. Kim, I.K. Oh, Theoretical and experimental investigation of the shape memory properties of an ionic polymer-metal composite, Smart Mater. Struct. 26 (4) (2017). Available from: https://doi.org/10.1088/1361-665X/aa61e9.

[34] Q. Shen, T. Stalbaum, N. Minaian, I.-K. Oh, K.J. Kim, A robotic multiple-shape-memory ionic polymer−metal composite (IPMC) actuator: modeling approach, Smart Mater. Struct. 28 (2019) 015009. Available from: https://doi.org/10.1088/1361-665X/aaeb83.

[35] W. Grot, Fluorinated Ionomers, second ed., Elsevier, Chadds Ford, PA, 2011.

[36] S. Trabia, Z. Olsen, K.J. Kim, Searching for a new ionomer for 3D printable ionic polymer-metal composites: aquivion as a candidate, Smart Mater. Struct. 26 (11) (2017). Available from: https://doi.org/10.1088/1361-665X/aa919f.

[37] M. Shahinpoor, K.J. Kim, Ionic Polymer Metal Composites (IPMCs), RSC, 2015, p. 1. Available from: https://doi.org/10.1039/9781782622581.

[38] M. Kotal, J. Kim, K.J. Kim, I.K. Oh, Sulfur and nitrogen co-doped graphene electrodes for high-performance ionic artificial muscles, Adv. Mater. 28 (8) (2016) 1610−1615. Available from: https://doi.org/10.1002/adma.201505243.

[39] H.S. Wang, J. Cho, D.S. Song, J.H. Jang, J.Y. Jho, J.H. Park, High-performance electroactive polymer actuators based on ultrathick ionic polymer-metal composites with nanodispersed metal electrodes, ACS Appl. Mater. Interfaces 9 (26) (2017) 21998−22005. Available from: https://doi.org/10.1021/acsami.7b04779.

[40] K. Bian, H. Liu, G. Tai, K. Zhu, K. Xiong, Enhanced actuation response of nafion-based ionic polymer metal composites by doping BaTiO3 nanoparticles, J. Phys. Chem. C. 120 (23) (2016) 12377–12384. Available from: https://doi.org/10.1021/acs.jpcc.6b03273.

[41] S. Ruiz, B. Mead, V. Palmre, K.J. Kim, W. Yim, A cylindrical ionic polymer-metal composite-based robotic catheter platform: modeling, design and control, Smart Mater. Struct. 24 (1) (2015). Available from: https://doi.org/10.1088/0964-1726/24/1/015007.

[42] S.J. Kim, D. Pugal, J. Wong, K.J. Kim, W. Yim, A bio-inspired multi degree of freedom actuator based on a novel cylindrical ionic polymer-metal composite material, Robot. Auton. Syst. 62 (1) (2014) 53–60. Available from: https://doi.org/10.1016/j.robot.2012.07.015.

[43] M. Shahinpoor, K.J. Kim, Effect of surface-electrode resistance on the performance of ionic polymer-metal composite (IPMC) artificial muscles, Smart Mater. Struct. 9 (4) (2000) 543–551. Available from: https://doi.org/10.1088/0964-1726/9/4/318.

[44] B. Lopes, P.J. Costa Branco, Ionic polymer metal-composite (IPMC) actuators: augmentation of their actuation force capability, in: IECON Proceedings (Industrial Electronics Conference), 2009, pp. 1180–1184. https://doi.org/10.1109/IECON.2009.5414659.

[45] P.B. Michael, H.L. Jeffrey, J. Li, A. Slocum, Optimum design of an electrostatic zipper actuator, in: 2004 NSTI Nanotechnology Conference and Trade Show - NSTI Nanotech 2004, 2004, p. 2.

[46] N. Kellaris, V.G. Venkata, P. Rothemund, C. Keplinger, An analytical model for the design of Peano-HASEL actuators with drastically improved performance, Extreme Mech. Lett. 29 (2019) 100449. Available from: https://doi.org/10.1016/j.eml.2019.100449.

[47] T. Park, Y. Cha, Soft Gripper Actuated Electro-Hydraulic Force, 2019, p. 46. https://doi.org/10.1117/12.2514008.

[48] C.A. Manion, M. Fuge, S. Bergbreiter, Modeling and evaluation of additive manufactured HASEL actuators*, in: International Conference on Intelligent Robots and Systems (IEEE/RSJ), 2018.

[49] Y. Li, M. Hashimoto, PVC gel soft actuator-based wearable assist wear for hip joint support during walking, Smart Mater. Struct. 26 (12) (2017). Available from: https://doi.org/10.1088/1361-665X/aa9315.

[50] K. Adesanya, E. Vanderleyden, A. Embrechts, P. Glazer, E. Mendes, P. Dubruel, Properties of electrically responsive hydrogels as a potential dynamic tool for biomedical applications, J. Appl. Polym. Sci. 131 (23) (2014)n/a-n/a. Available from: https://doi.org/10.1002/app.41195.

[51] S. Murdan, Electro-responsive drug delivery from hydrogels, J. Control. Release 92 (2003) 1–17.

[52] G. Kim, H. Kim, I.J. Kim, J.R. Kim, J.I. Lee, M. Ree, Bacterial adhesion, cell adhesion and biocompatibility of Nafion films, J. Biomater. Sci. Polym. Ed. 20 (12) (2009) 1687–1707. Available from: 10.1163/156856208X386273.

[53] L. Sutton, H. Moein, A. Rafiee, J.D.W. Madden, C. Menon, Design of an assistive wrist orthosis using conductive nylon actuators, Proceedings of the IEEE RAS and EMBS International Conference on Biomedical Robotics and Biomechatronics, IEEE, 2016, pp. 1074–1079. Available from: https://doi.org/10.1109/BIOROB.2016.7523774.

CHAPTER 4

An optimized soft actuator based on the interaction between an electromagnetic coil and a permanent magnet

Nafiseh Ebrahimi[1], Paul Schimpf[2] and Amir Jafari[1]

[1]Advanced Robotic Manipulators ARM Lab, Department of Mechanical Engineering, College of Engineering, University of Texas at San Antonio UTSA, San Antonio, TX, United States
[2]Department of Computer Science, Eastern Washington University (EWU), Cheney, WA, United States

4.1 Introduction

Electromagnetic actuators are gaining interests among researchers due to many advantages including short response time, simple controllability, an uncomplicated structure in comparison to the other types of the actuator [1,2].

The design, optimization, and application of the solenoid as a vastly used electromagnetic actuator has been considered by many researchers in recent years [3]. A solenoid is composed of a coil that is an electrically conductive wire warped around a magnetically permeable cylindrical core, that is, plunger. When an electric current passes through the coil, it generates a magnetic field that can act upon the plunger and create electromagnetic force. This type of linear actuators has served numerous applications ranging from measurement systems to manufacturing. Banick and Haller [4] designed a solenoid actuator equipped with a magnetic flux sensor capable of indicating the position of the solenoid core. Lim et al. [5] proposed a method for proportional control of a solenoid actuator to convert its switching mode into a proportional actuator. Mitsutake et al. [6] applied finite element method to predict the dynamic response characteristics of a linear solenoid actuator. Kamal and Daehn [7] presented an analysis for the coil to design an electromagnetic actuator for flat sheet forming purposes. Petit et al. [1] proposed a four-discrete-position electromagnetic actuator, then presented its modeling and also experiments. Shin et al. [8] developed a biomimetic actuator using four segmented solenoids mimicking earthworm movements. Fries et al. [9] fabricated an electromagnetically driven elastic actuator proper for use as muscle-like structures, capable of generating stress and strain when embedded in a solenoid coil. Song and Lee [2] developed a solenoid actuator with a ferromagnetic plunger and applied the solenoid actuator in a multisegmented miniaturized robot to generate both rectilinear and turning movements. Said et al. [10] designed and fabricated a compact electromagnetically driven microactuator using polydimethylsiloxane (PDMS) and embedded magnetic particles. Rawlik et al. [11] presented a method to design a coil generating an arbitrarily shaped magnetic field. Guo et al. [12] demonstrated the fabrication of a soft electromagnetic actuator using a liquid metal coil of Ga-In alloy for soft robotic applications. Nevertheless, there is still a deficiency in coil design optimization in order to generate the optimum possible field and force out of the solenoid actuator.

Previously, we discussed the design, development, and control of an electromagnetic soft actuator for rehabilitation application that comprises two solenoid coils and a common magnetic core. This actuator is entirely made of soft materials and is highly scalable [13]. The actuator body is made of silicone polymer PDMS with embedded helical microchannels. Eutectic gallium–indium (EGaIn) that is a conductive liquid in room temperature was injected into microchannels creating conductive coils. Magnetic particles are mixed with PDMS and placed into a strong magnetic field during the curing process to force magnetic particles to stay aligned together once the external magnetic field was removed. The result is a soft permanent magnet which was placed into the solenoids as the core. In this paper, we propose an electromagnetic actuator, presented in Fig. 4.1, consisting of two coils and a permanent magnet plunger in the middle interacting with the coils to generate linear motion and force. The magnetic plunger was applied in order to generate higher actuation forces compared to the ferromagnetic core [2]. Moreover, this magnetic plunger is shared between two coils to benefit the entire volume of the magnet to raise the generated force compared to the generic solenoid actuators.

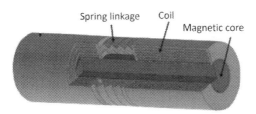

Figure 4.1 Schematic of electromagnetic actuator.

We intended to find a proper configuration for the components as well as solenoid design characteristics in order to exert maximum output force out of the actuator. To achieve this goal, we implemented our theoretical analysis using the Biot–Savart law [14] to calculate the magnetic field of the solenoid on the axis. Then, the applied force of the solenoid upon the permanent magnet plunger and the total applied force of the electromagnetic actuator were derived utilizing the charge model [15,16]. Then, the geometrical design of the solenoid and plunger was considered in the next steps by applying the analytical results for the magnetic field and force. The influence of cylindrical cross section deformation on the magnetic field at the center of the solenoid was investigated as well.

4.2 Solenoid magnetic field and force calculation

Our proposed actuator is composed of two electromagnetically inductive solenoids combined with permanent magnet plungers. The force generated by the solenoid depends on several factors including the number of turns, the coil's length and diameter, applied current, and the permanent magnet's length, diameter, and material.

To design the electromagnetic actuator, a theoretical analysis was conducted on the electromagnetic force produced by two solenoids encompassing a common permanent magnetic core. For this purpose, firstly the

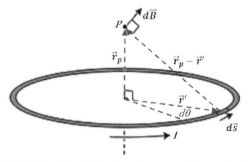

Figure 4.2 Magnetic field at the desired point caused by a single circular current element. $d\vec{B}$, Differential element of magnetic field at the center of the current element; $\vec{r}_p$, axial position vector of the arbitrary point P; $\vec{r}$, radial position vector of the arbitrary point P (radius vector); I, flowing current; $d\vec{s}$, differential length element on the circular current carrying loop.

$$d\vec{B} = \frac{\mu_0 I}{4\pi} \frac{d\vec{s} \times \hat{r}}{r^2} = \frac{\mu_0 I}{4\pi} \frac{d\vec{s} \times \vec{r}}{r^3} = \frac{\mu_0 I}{4\pi} \frac{d\vec{s} \times (\vec{r}_p - \vec{r}')}{|\vec{r}_p - \vec{r}'|^3} \qquad (4.1)$$

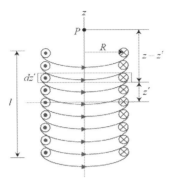

Figure 4.3 Magnetic field at a desired point generated by a schematic coil. L, Solenoid length; R, solenoid radius; z, axial distance of the desired point P to the center of the solenoid and along its axis; dz′, differential current element thickness; z′, axial distance of current element to the center of the solenoid.

Biot–Savart law calculates the magnetic field along the axis of a loop passing steady current at the arbitrary point P. Applying the Biot–Savart law, the contribution of the current element (Fig. 4.2) to the magnetic field at point P is expressed by Eq. (4.1) [14,17]: where $d\vec{s} = Rd\theta\hat{\theta}$, $\vec{r}_p = z\hat{k}$, $\vec{r}' = R\hat{r}$, and R is the magnitude of the radius vector ($\vec{r}'$) or coil radius shown in Fig. 4.3. Also, z is the coordinate along the solenoid axis and $\hat{k}$ is the unit vector in z direction.

$$d\vec{B} = \frac{\mu_0 I (Rd\theta\hat{\theta}) \times (z\hat{k} - R\hat{r})}{4\pi |z\hat{k} - R\hat{r}|^3}$$

$$= \frac{\mu_0 IzR}{4\pi(z^2 + R^2)^{3/2}} d\theta \hat{r} + \frac{\mu_0 IR^2}{4\pi(z^2 + R^2)^{3/2}} d\theta \hat{k} \quad (4.2)$$

Integrating over the entire circular loop results in the magnetic field at P. The first integral vanishes due to the equal magnitude and the fact that radial unit vectors around the circle sum to zero. Hence, that remaining is just the axial component of the magnetic field Eq. (4.3).

$$\vec{B}_Z = \frac{\mu_0 IR^2}{4\pi(z^2 + R^2)^{3/2}} \int_0^{2\pi} d\theta \hat{k} = \frac{\mu_0 IR^2}{2(z^2 + R^2)^{3/2}} \hat{k} \quad (4.3)$$

For a finite solenoid consisting of a large number of circular loops using the result obtained above for the magnetic field of one loop, the magnetic field at point P on the axis of the solenoid could be obtained by integrating over the entire length. Fig. 4.3 shows the selected current element on the solenoid. The z dimension of the selected point (P) is always measured from the central loop of the solenoid. The amount of current passing through the element is given by

$$dI = I(ndz') = I\left(\frac{N}{l}\right)dz',$$

where $n = N/l$ is the number of turns per unit length or turn density.

Applying Eq. (4.3), the contribution to the magnetic field at P caused by the current element with a thickness of dz' is:

$$dB_z = \frac{\mu_0 R^2}{2[(z-z')^2 + R^2]^{3/2}} dI = \frac{\mu_0 R^2}{2[(z-z')^2 + R^2]^{3/2}} (nIdz') \quad (4.4)$$

Integrating from the Eq. (4.4) over the solenoid length, we obtain:

$$\begin{aligned}
B_z &= \frac{\mu_0 nIR^2}{2} \int_{-l/2}^{l/2} \frac{dz'}{[(z-z')^2 + R^2]^{3/2}} \\
&= \frac{\mu_0 nIR^2}{2} \frac{z'-z}{R^2\sqrt{(z-z')^2 + R^2}} \Big|_{-l/2}^{l/2} \\
&= \frac{\mu_0 nI}{2} \left[\frac{(l/2)-z}{\sqrt{(z-l/2)^2 + R^2}} + \frac{(l/2)+z}{\sqrt{(z+l/2)^2 + R^2}}\right]
\end{aligned} \quad (4.5)$$

Eq. (4.5) expresses the magnetic field of a finite solenoid at the desired point P with a distance of z from the center of the solenoid. For our electromagnetic actuator system, we need to take into account the contribution of both coils on the particular point in which we are interested to find the magnetic field. Using the result obtained above for a single solenoid and applying the superposition principle the magnetic field at the desired point on the common axis of two solenoids (P) is obtainable.

$$\mathbf{B}_{ext} = \mathbf{B}_{right} + \mathbf{B}_{left} \quad (4.6)$$

To calculate $\mathbf{B}_{right}$ and $\mathbf{B}_{left}$ in Eq. (4.6), the distances of an arbitrary point P from the center of each coil (z_1 & z_2) should be determined as depicted in Fig. 4.4.

Next step is the calculation of applied force to the magnetic core. Charge model is a useful method for analyzing a permanent magnet, such as determining the force and torque on a magnet located inside an

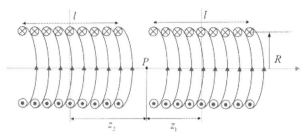

Figure 4.4 Magnetic field calculation at point P on the common axis of two schematic coaxial solenoids of the actuator. L, Solenoid length; R, solenoid radius; z_1, axial distance of the arbitrary point P to the center of the right-hand solenoid; z_2, axial distance of the arbitrary point P to the center of the left-hand solenoid.

external field. Based on the charge model force can be expressed by Eq. (4.7) [16]:

$$\mathbf{F} = \int_v \rho_m \mathbf{B}_{ext} dv + \oint_s \sigma_m \mathbf{B}_{ext} ds \tag{4.7}$$

where $\rho_m = -\nabla \cdot \mathbf{M}(Amps/m^2)$ is equivalent volume charge density and $\sigma_m = \mathbf{M} \cdot \hat{n}(Amps/m)$ is equivalent surface charge density

In this case, $\mathbf{B}_{ext}$ is the superposed field calculated from Eqs. (4.5) and (4.6) at any arbitrary point ($\mathbf{B}_p$). The magnetic core has a fixed and uniform magnetization along its axis expressed as follows:

$$\mathbf{M} = M\hat{\mathbf{Z}}$$

Therefore, $\rho_m = -\nabla \cdot \mathbf{M} = 0$.

The magnetization, $\mathbf{M}$, is the net magnetic moment per unit volume of the permanent magnet and can be calculated using Eq. (4.8):

$$M = \frac{B_r}{\mu_0} \tag{4.8}$$

where μ_0 is the vacuum permeability and B_r is a material property named remanence or residual flux density. It is an important factor for permanent magnets and is obtainable from the hysteresis curve of the material [18].

To calculate surface charge (σ_m) we first should determine the unit surface normal vectors on the magnet. The cylindrical magnet has three distinct surfaces as follows (Fig. 4.5).where h and r are the magnet core length and radius respectively. Also, $\hat{z}$ and $\hat{r}$ are unit vectors in axial and radial directions.

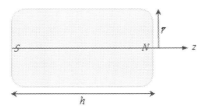

Figure 4.5 Magnet core dimensions. $\bar{r}$, Permanent magnet radius, h, permanent magnet length.

$$\hat{n} = \begin{cases} -\hat{\mathbf{z}} & z=0 \\ \hat{\mathbf{z}} & z=h \\ \hat{\mathbf{r}} & r=\bar{r} \end{cases}$$

Then, the surface charge density $\sigma_m = \mathbf{M} \cdot \hat{n}$ is obtained as:

$$\sigma_m = \begin{cases} -M_s & z=0 \\ M_s & z=h \end{cases} \tag{4.9}$$

where M_s is the magnetization of the magnetic core along the z axis. It is notable that for the cylindrical surface of the magnet, $\sigma_m = \mathbf{M} \cdot \hat{n} = 0$ since the two vectors are perpendicular to each other. The north pole surface of the magnetic core is located in the right solenoid and the south pole surface is located in the left one (Fig. 4.6). Dimension d in this figure expresses the distance between the magnet pole and the related coil middle point.

The magnetic core is located exactly at the middle of the line connecting the two coils, so each pole has the same distance to the correlated coil's end (d). Due to the symmetric geometry of the actuator, we just need to calculate the force applied to one end surface of the magnetic core and then double it to obtain the total applied force. These two forces are identical in magnitude but have opposite directions. If we consider the north pole of the magnet, firstly we need to calculate $\mathbf{B}_{ext}$ at this point. Choosing an arbitrary point, P, on the north pole z_1 and z_2 could be calculated as: $z_1 = d$ and $z_2 = d + h$.

Applying Eqs. (4.5) and (4.7) the contributions of both coils at this specific point could be calculated and superposed.

Having the same current directions in two coils results in two same direction magnetic field vectors algebraically added together presented by Eq. (4.11).

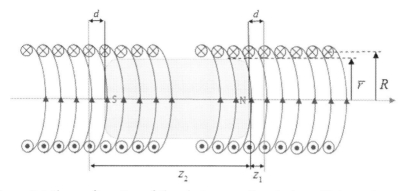

Figure 4.6 The configuration of the electromagnetic actuator with two schematic coils and common magnetic core. R, Solenoid radius; r̄, permanent magnet radius; d, axial distance from each magnet pole to the center of related solenoid; z_1, distance between north pole and the center of the right solenoid; z_2, distance between the north pole and the center of the left solenoid.

$$\mathbf{B}_{right} = \frac{\mu_0 nI}{2}\left[\frac{(l/2)+d}{\sqrt{(-d-(l/2))^2+R^2}} + \frac{(l/2)-d}{\sqrt{(-d+(l/2))^2+R^2}}\right] \quad (4.10)$$

$$\mathbf{B}_{left} = \frac{\mu_0 nI}{2}\left[\frac{(l/2)-(d+h)}{\sqrt{(d+h-(l/2))^2+R^2}} + \frac{(l/2)+(d+h)}{\sqrt{(d+h+(l/2))^2+R^2}}\right]$$

$$B_{ext} = B_{right} + B_{left}$$

$$= \frac{\mu_0 nI}{2}\left[\begin{array}{c}\dfrac{(l/2)+d}{\sqrt{(-d-(l/2))^2+R^2}} + \dfrac{(l/2)-d}{\sqrt{(-d+(l/2))^2+R^2}} \\ + \dfrac{(l/2)-(d+h)}{\sqrt{(d+h-(l/2))^2+R^2}} + \dfrac{(l/2)+(d+h)}{\sqrt{(d+h+(l/2))^2+R^2}}\end{array}\right] \quad (4.11)$$

Hence, applying Eq. (4.8), the force on the magnetic core is obtained:

$$\mathbf{F} = \oint_s \sigma_m \mathbf{B}_{ext} ds = \mathbf{B}_{ext} M \int_0^{\bar{r}}\int_0^{2\pi} r\,dr\,d\phi = \mathbf{B}_{ext}\frac{B_r}{\mu_0}\pi\bar{r}^2 \quad (4.12)$$

Substituting Eq. (4.11) into (4.12) results in a force applied to the north pole of the magnetic core, generated by two right and left coils' magnetic fields. This force is presented by Eq. (4.13).

$$F = \frac{nIB_r}{2}\pi\bar{r}^2 \left[\frac{(l/2)+d}{\sqrt{(d+(l/2))^2 + R^2}} + \frac{(l/2)-d}{\sqrt{(-d+(l/2))^2 + R^2}} + \frac{(l/2)-(d+h)}{\sqrt{(d+h-(l/2))^2 + R^2}} + \frac{(l/2)+(d+h)}{\sqrt{(d+h+(l/2))^2 + R^2}} \right]$$

(4.13)

The same story goes true for the south pole. Regarding the symmetry, the south pole experiences the same amount of force on the correlated surface. Hence, Eq. (4.14) expresses the whole amount of force applied to the magnet core:

$$F_{total} = 2F$$

$$= nIB_r\pi\bar{r}^2 \left[\frac{(l/2)+d}{\sqrt{(d+(l/2))^2 + R^2}} + \frac{(l/2)-d}{\sqrt{(-d+(l/2))^2 + R^2}} + \frac{(l/2)-(d+h)}{\sqrt{(d+h-(l/2))^2 + R^2}} + \frac{(l/2)+(d+h)}{\sqrt{(d+h+(l/2))^2 + R^2}} \right]$$

(4.14)

The maximum magnetic field and consequently, solenoid force occurs at the center of the coil. Hence, the centers of the coils would be the best place for positioning the permanent magnet base surfaces or poles ($d = 0$ in Eqs. (4.11) and (4.13)). Employing the two equations, Fig. 4.7A and B depict such an actuator net magnetic field and force variations, respectively, in terms of inner diameter (d_i). In this analysis, some of the parameters get fixed as follows: the coil wire length ($l_w = 3$ m), the applied current to the actuator coils ($i = 0.33$ amp) based on [19], the permanent magnet length ($h = 10$ mm), the radial air-gap between magnet core and coils ($g = 0.2$ mm), magnet core residual flux density ($B_r = 0.18$ T), and the wire gauge (34AWG). We tried to find the optimum amount of inner diameter for a one-layer coil made by a fixed wire length ($l_w = 3$ m).

Fig. 4.7A shows a maximum magnetic field at 4 mm. When the inner diameter is larger than almost 4 mm, the magnetic field decreases due to the increase in distance vector ($\vec{r}$) length. On the other hand, since the total force is proportional to the square of the inner diameter, it eventually grows as the inner diameter rises (Fig. 4.7B). Since the wire length and applied current are constant, all different coil combinations consume equal electrical powers.

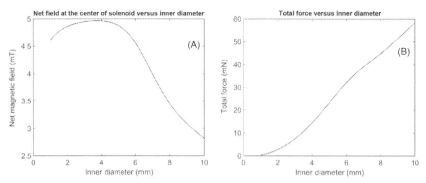

Figure 4.7 (A) Net magnetic field at the center of each coil of the actuator versus coil inner diameter (l_w, wire length = 3 m, permanent magnet length = 10 mm). (B) The total force of the actuator versus coil inner diameter. (l_w, wire length = 3 m, permanent magnet length = 10 mm).

4.3 Solenoid geometry design optimization

In this section, the determination of optimal values for coil geometry is discussed in order to obtain maximum magnetic field strength at the center of the coil bounding the electrical power consumed by the coil. We restrict the solenoid's wire length (l_w) affecting the resistance and also flowing current (0.33 amp) through the conductor to ensure a certain amount of power consumption. The geometrical properties of a coil include the solenoid length (l), the inner and outer diameter (d_i, d_o, respectively), and the number of coil turns (N). According to the previous section's analytical results (Fig. 4.7B) the solenoid force has an ascending relationship with the solenoid inner diameter. However, the magnetic field reaches the maximum nearly 4 mm inner diameter, which is determined for the optimization analysis. In order to find optimal solenoid length, we studied the distribution of certain wire length (l_w = 3 m) over different coil lengths, varying from 0.16 mm (the wire diameter) in a flat spiral coil to 42 mm in a one-layer long coil. We performed this on a wide range of coil turn distributions from the multilayer planar coil all the way to the one-layer one just to get the idea as to how the optimal geometry would be look like. Applying $z = 0$ in Eq. (4.5), which means selecting the arbitrary point (P) at the central loop, gives the maximum magnitude of the magnetic field at the center of the solenoid (Fig. 4.8A). Also, the force resulting from the interaction of the coil and the magnetic core is presented in Fig. 4.8B.

The final goal of this section is to obtain an optimal multilayer geometry for the coil, therefore we need to know how various geometrical parameters

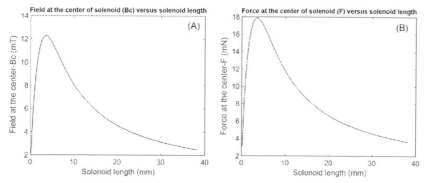

Figure 4.8 (A) The magnetic field at the center of solenoid versus the solenoid length for a fixed length wire ($l_w = 3$ m). (B) Force at the center of solenoid versus the solenoid length for a fixed length wire ($l_w = 3$ m).

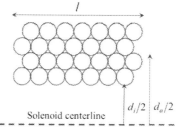

Figure 4.9 Coil axial cross section and packing density (λ). $d_i/2$, Coil inner radius; $d_a/2$, coil average radius; l, coil length.

of a coil are correlated to each other. It is worth noting that, the radius in Eq. (4.5) must be the coil average radius ($d_a/2$) which is the function of turn numbers (N) and circle packing density (λ). Circle packing density is the ratio of the cross section taken up by the wires to the available space. The maximum packing density is for the hexagonal lattice circle packing arrangement shown in Fig. 4.9 which has a packing density of $\pi/\sqrt{12}$ [20,21].

The relationship among coil average and internal diameters (d_a, d_i), coil length (l), packing density (λ), turn numbers (N), and wire cross section (a) is explained in detail in [20], as follows in Eq. (4.15):

$$\lambda(d_a - d_i)l = Na$$
$$d_a = \frac{a}{\lambda l}N + d_i \qquad (4.15)$$

Moreover, N is also considered a function of average diameter (d_a) and wire length (l_w) by itself, as expressed in Eq. (4.16).

Therefore the turn density ($n = N/l$) is a function of wire length and average diameter.

$$N = \frac{l_w}{\pi d_a} \qquad (4.16)$$

Fig. 4.10 depicts the variation of the solenoid's average and external diameter with respect to solenoid length. The average diameter falls significantly as the solenoid length grows.

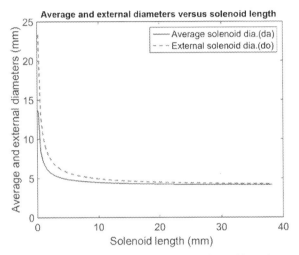

Figure 4.10 Average and external diameters versus solenoid length.

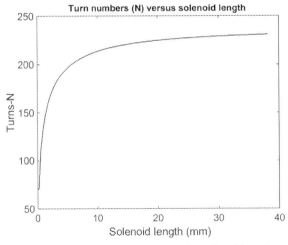

Figure 4.11 Solenoid turns versus solenoid length for a fixed length wire ($l_w = 3$ m).

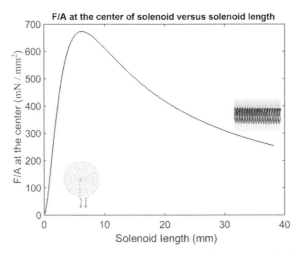

Figure 4.12 Force to solenoid cross section ratio at the center of solenoid versus solenoid length for a fixed length wire ($l_w = 3$ m).

Fig. 4.11 illustrates the changes in solenoid turns (N) versus solenoid length. The increase in N is due to the drastic decrease in average solenoid diameter while the solenoid length rises.

In all coil configurations investigated above, the wire length and flowing current were kept identical, so, the electrical power consumption for all of them is $P = RI^2 = 0.2797$ watts.

The goal is to gain the maximum force out of a certain cross-sectional area of the solenoid. Hence, we maximized the ratio of force to cross section (F/A), where A is the cross section of the solenoid calculated by the external coil diameter (d_o). Fig. 4.14 illustrates how the (F/A) ratio varies with respect to the solenoid length.

The optimal amount for the solenoid length according to the graph in Fig. 4.12 is $l = 6.4$ mm which maximizes the force to cross section ratio to 662 mN/mm^2. The corresponding turn numbers, average coil diameter, and external coil diameter at this optimum length were obtained from graphs in Figs. 4.10 and 4.11 as $d_a = 4.7$ mm, $d_o = 5.4$ mm, and $N \approx 215$ turns, respectively.

The aforementioned solenoid optimal geometry also maximizes the (F/A) ratio drawn versus the solenoid cross section in Fig. 4.15.

Conducting the analytical optimization above we found an optimum solenoid geometry for embedding in the electromagnetic actuator structure to maximize the exerting force out of the solenoid size while bounding the power consumption to a certain amount. The optimal coil geometry

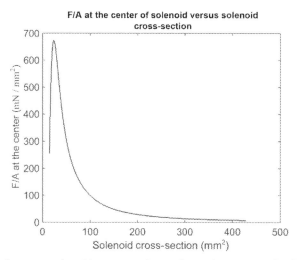

Figure 4.13 Force to solenoid cross section ratio at the center of solenoid versus solenoid cross section for a fixed length wire ($l_w = 3$ m).

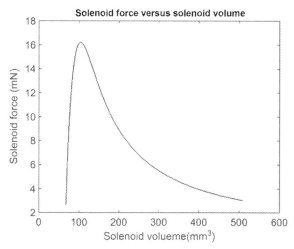

Figure 4.14 Force at the center of solenoid versus the solenoid volume for a fixed length wire ($l_w = 3$ m).

comprised almost 215 turns in 40 columns and nearly five layers. The coil inner and outer diameters are $d_i = 4$ mm and $d_o = 5.4$ mm. According to the graphs presented in Fig. 4.8A and B, the maximum amounts for the magnetic field and force at the center of the designed optimal coil are $B_{max} = 12.26$ mT and $F_{max} = 17.88$ mN, respectively.

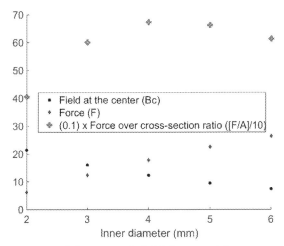

Figure 4.15 Comparison of the magnetic field, force, and force to cross section ratio for different coil inner diameters (2, 3, 4, 5, and 6 mm).

Figs. 4.12 and 4.13 illustrate that increasing the size of the solenoid actuator leads to a drastic decrease in the actuation force. The diagrams have an ascending trend on the small beginning portion of the diagram where the solenoid sizes are smaller than the optimum one. From the peak point onwards, the actuation force decreases by enlarging the solenoid. In other words, the proposed solenoid actuator is scalable and has generally the capability of generating higher forces by getting smaller. Fig. 4.14, in addition, strongly confirms this finding as it has almost the same behavior since the solenoid force drops significantly by increasing the volume of the solenoid. This characteristic enables one to exert higher forces by combining several tiny actuators rather than using just a larger one. However, there are some limitations in scaling down the size of the actuator which are discussed in the following section.

In Fig. 4.15 we compared various coil inner diameters in terms of three parameters: magnetic field, force, and force to cross section ratio. It verifies that our selection for inner diameter, 4 mm, generates the maximum F/A among the other options.

4.4 Manufacturing aspects and limitations

In this section, the restrictions of scaling down the size of the actuator are discussed. In the previous section, we just selected a constant wire length and diameter to bound the power consumption and optimize the geometry of the coil to generate higher forces. Regarding such constraints, we found some relatively small dimensions for the solenoid which optimize the generated force. It was argued that enlarging those dimensions causes less force to cross section or force to volume ratios. Although the applied constraints over the wire length and its diameter originate in some manufacturing and implementation restrictions, we studied various wire lengths as well as wire diameters to show how they affect the generated force and power consumption.

Firstly, we considered coils with different wire length with the same diameter (34AWG), as presented in Table 4.1. The results show that longer wires consume more power, generating larger coils which produce more force to size ratio. However, the last column of the table compares the generated force over the size and consumed power. Comparing the last column of Table 4.1, as expected, the smaller coil made up of the shorter wire is the most effective option. Despite the efficiency of the mentioned coil, it is not executable thanks to its small size ($d_i = 1.85$ mm). Therefore the wire length of 3 m was chosen for the optimization study.

Secondly, we studied coils composed of equal wire length ($l_w = 3$ m) but different diameters. Since the wire diameters are different, the allowable applying current [19] varies in each case as well as the resistance per length [22], both presented in Table 4.2. In this case again the finest wire ($40AWG$) generates the highest amount of force to cross section to power ratio (Table 4.2). Even though, due to the very fine geometry of the correlated coil ($d_i = 2.27$ mm) and small allowable current ($I = 0.09$ A), it is not applicable. Consequently, the wire size 34AWG was selected.

Table 4.1 Comparison of coils with various wire lengths but the same wire diameter (34AWG).

Wire length (l_w, m)	Optimum inner dia. (di) mm	Optimum coil length (l, mm)	Consumed power (P, watt)	Force to cross section ratio (F/A)	Force to cross section to power ratio (F/AP)
1	1.85	5.29	0.0932	325	3.48×10^3
2	2.98	5.85	0.1864	541	2.90×10^3
3	3.69	6.44	0.2797	662	2.37×10^3
4	4.39	6.91	0.3729	756	2.03×10^3
5	4.82	7.39	0.4661	827	1.78×10^3

Table 4.2 Comparison of coils with various wire dia. but the same wire length ($l_w = 3$ m).

Wire gauge	Wire dia. (d_w, mm)	Resistance/length (mΩ/m)	Allowable current (A)	Optimum inner dia. (d_i, mm)	Optimum coil length (l, mm)	Force to cross section to power ratio (F/AP)	Force to cross section ratio (F/A)
34AWG	0.160	856	0.33	3.7	6.4	2.36	0.66
36AWG	0.127	1361	0.21	3.1	5.6	3.04	0.55
38AWG	0.101	2164	0.13	2.7	4.8	3.99	0.44
40AWG	0.0799	3441	0.09	2.3	4.1	4.65	0.39

Hence, there are some factors that contribute to practical manufacturing and experimental conditions restricting our selection of the optimum coil for the electromagnetic actuator.

4.5 The influence of solenoid section deformation on the magnetic field and force

The actuator might undergo some squeezing and loses its round shape under the transverse loads. In this section, we are going to study the compressive effect on the resultant magnetic field. Under radial load, the circular cross section turns into the ellipse and then, by increasing the load eccentricity of the elliptical cross section, will increase. In the following we consider the resulting magnetic field of a planar elliptical current carrying conductor at the center of the geometry and then investigate the effect of ellipse eccentricity on the field. For this purpose, we consider a current carrying elliptical wire geometry and then again use the Biot–Savart law to explicitly calculate the magnetic field due to the flowing current. This law gives the total magnetic field at an arbitrary point P in the space by superposition of magnetic field contributors. Having vectorial form, only a few simple conductor geometries lead to an analytical evaluation of the magnetic field and shapes other than a circular or straight wire lead to some complicated integral calculations [23]. Applying polar coordinates for a general equation of conic sections (including ellipse) and using a vectorial form of the Biot–Savart law, an equation for the magnetic field has been derived in [24,25]. However, it is valid just at a focus of the sections.

Using the polar coordinate system for the expression of the wire's geometry, Miranda [23] proposed a very simple line integral (Eq. 4.17) using the Biot–Savart law in scalar form in order to calculate the magnitude of the magnetic field due to arbitrary shapes of planar current-carrying conductors at a point belonging to the plane of the wire.

$$B = \frac{\mu_0 I}{4\pi} \oint \frac{d\theta}{r} \qquad (4.17)$$

where $r = r(\theta)$ is the expression of wire geometry in polar coordinate and measured from the observation point (O) located in the same plane with the wire [23,26,27]. For an elliptical wire geometry shown in Fig. 4.16,

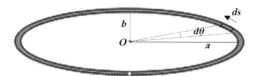

Figure 4.16 Current element on the elliptical conductor loop. a, Semimajor axis of the ellipse; b, semiminor axis of the ellipse, differential length element on the elliptical current carrying loop.

the element of ds normal to r is $rd(\theta)$. In addition, $x = r\cos\theta$ and $y = r\sin\theta$ so that the line integral above expresses as the following:

$$B = \frac{\mu_0 I}{4\pi} \oint \frac{d\theta}{r} = \frac{\mu_0 I}{4\pi} \oint \frac{\sqrt{a^2 \sin^2\theta + b^2 \cos^2\theta} d\theta}{ab} \tag{4.18}$$

Since, for ellipse geometry $\frac{x^2}{a^2} + \frac{y^2}{b^2} = 1$, $1 = \frac{r^2 \cos^2\theta}{a^2} + \frac{r^2 \sin^2\theta}{b^2} = r^2 a^{-2} b^{-2} (b^2 \cos^2\theta + a^2 \sin^2\theta)$.

On the other hand, the area of the ellipse is

$$A = \pi ab \tag{4.19}$$

And the perimeter is

$$\oint \sqrt{dx^2 + dy^2} = \oint \sqrt{a^2 \sin^2\varphi + b^2 \cos^2\varphi} d\varphi = P \tag{4.20}$$

Since, the parametric form of the ellipse equation is $x = a\cos\varphi$, $y = b\sin\varphi$; $dx = -a\sin\varphi d\varphi$, and $dy = a\cos\varphi d\varphi$. Substituting Eqs. (4.19) and (4.20) in Eq. (4.18), results in:

$$B = \frac{1}{4} \mu_0 IP/A \tag{4.21}$$

So, the magnitude of the magnetic field at the center of an elliptical conductor is proportional to the ratio of the circumference to the area.

We were interested to find the impact of solenoid section deformation on the magnetic field at the center. Considering Eq. (4.21), for the magnetic field of a one-loop circular section with specific perimeter we studied the effect of section deformation as a function of ellipse eccentricity ($0 \leq e = c/a < 1$) as the circular section undergoes deformation and turns to an ellipse. The parameter "c" is the distance from the center to a focus of the ellipse. Fig. 4.17 depicts the variation of magnetic field versus elliptical eccentricity beginning from $e = 0$ for a round circular coil section and growing to $e < 1$ for a squeezed oval coil.

130 Soft Robotics in Rehabilitation

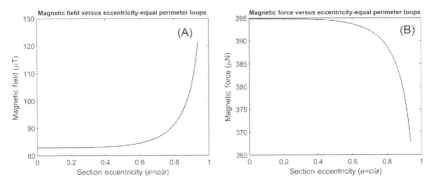

Figure 4.17 (A) Magnetic field changes due to deformation of one current carrying loop section. (B) Magnetic force changes due to deformation of one current carrying loop section.

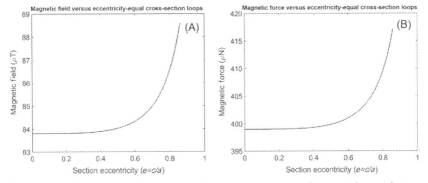

Figure 4.18 (A) Magnetic fields of elliptical current carrying loops with equal cross sections but different perimeters and eccentricities. (B) Magnetic forces of elliptical current carrying loops with equal cross sections but different perimeters and eccentricities.

Fig. 4.17A depicts a drastic growth in the magnetic field at the center of a current carrying loop at $e = 0.8$ showing that the compressed and oval cross section of a solenoid escalates the resultant magnetic field at the center. However, because of the decrease in cross section of the squeezed and oval cross section the magnetic core cross section also decreases leading to a significant drop in magnetic force, as shown in Fig. 4.17B.

Maintaining a constant cross section, we are interested in comparing the magnetic field and force of circular and elliptical shapes, respectively. Fig. 4.18A and B compare axial magnetic fields and forces for elliptical current carrying loops with the same cross section areas but different perimeters and eccentricity. As depicted in the graphs, both of the parameters show significant rises around $e = 0.8$. In this case, because of the constant cross sections, the force has also an ascending trend similar to the magnetic field.

4.6 Discussion and conclusion

This article discusses design optimization of a cylindrical solenoid with a permanent magnet core as the main component of the electromagnetic actuator. The magnetic field and resultant force of the solenoid are formulized using the Biot—Savart law and the charge model, respectively. Eqs. (4.5) and (4.14) denote the geometrical and physical parameters contributing to the magnetic field and force, respectively. The impact of cylindrical coil inner diameter (d_i) on the magnetic field and force was examined for a constant wire length.

Fig. 4.7A and B show the dependency of the magnetic field and force on the inner diameter for a constant current passing throughout the coil. Since the assumption of fixed wire length is considered, the average diameter decreases as the coil length increases. Therefore the number of turns increases, up to a point where a single layer is reached. Results show that an increase in turn numbers leads to an increase in both the magnetic field and force. We obtained an optimal value for the coil inner diameter which maximized the magnetic field and later used this value to optimize the geometry so that it maximizes the force to the solenoid cross section ratio. As the inner diameter increases, the force is enhanced due to the increased magnetic core cross section.

According to the mentioned calculation for magnetic field and force, a solenoid geometrical design optimization was discussed. For this study, a coil with a constant inner diameter and varying length was investigated. The magnetic field at the middle of the coil was maximized while the electrical power consumption was restricted by means of using a particular wire length with a specific wire cross section and also applying a constant current in the calculations. The applied current was chosen with regard to the wire gauge current restrictions. Using packing density factor and applying average diameter (d_a) in the relevant equations, the magnetic field at the middle of the coil shows a maximum around a certain coil length. Hence, we got an optimized geometry of the coil in order to get the maximum magnetic field in the middle while its power consumption is restricted.

We then investigated the effect of scaling down the coil size and surprisingly found that the smaller the coil's size the greater the force to cross section ratio for the constant consumed power situation, as well as the greater force to cross section to power ratio in the case that power consumption is not restricted, as illustrated in Figs. 4.13 and 4.14, Tables 4.1 and 4.2.

Next, since the coil might undergo some deformity due to the lateral forces, we investigated the effect of solenoid cross section compression turning the section from circle to oval shapes with different degrees of ellipticity. For this purpose, we used a simplified form of Biot–Savart law toward the calculation of the magnetic field at the center of the elliptic geometry and then formulized it for various ellipses with the same circumference yet various cross sections. Keeping constant perimeter (P) for the coil section, the growth in elliptical eccentricity causes a decrease in the cross section area (A) and thus raises the resultant magnetic field at the center since it is proportional to the ratio of ellipse perimeter to its cross section. The coil cross section deformation into an oval increases the magnetic field at the center, as illustrated in Fig. 4.17. The graph shows the dependency of the magnetic field at the center of a constant perimeter geometry varies from a circle to squeezed ellipses. It initiates a significant escalation around the eccentricity of $e = 0.8$. However, section squeezing decreases the cross section causing a sharp drop in the resultant force. Studies of elliptical sections with the same cross section areas but different perimeters and eccentricities shows that there are significant increases in both magnetic field and forces at the center of the current carrying loop as section eccentricity increases.

Acknowledgment

This work is supported by the National Science Foundation under Grant No. 1850898.

References

[1] L. Petit, C. Prelle, E. DorÉ, F. Lamarque, M. Bigerelle, A four-discrete-position electromagnetic actuator: modeling and experimentation, IEEE/ASME Trans. Mechatron. 15 (1) (2010) 88–96. Available from: https://doi.org/10.1109/TMECH.2009.2017018.
[2] C.-W. Song, S.-Y. Lee, Design of a solenoid actuator with a magnetic plunger for miniaturized segment robots, Appl. Sci. 5 (3) (2015) 595–607. Available from: https://doi.org/10.3390/app5030595.
[3] Zahra Abbasi, Jesse Hoagg, Thomas Seigler, Decentralized Position and Attitude Control for Electromagnetic Formation Flight, AIAA Scitech 2019 Forum: Guidance, Navigation, and Control (2019). Available from: https://doi.org/10.2514/6.2019-0908.

[4] G.S. Banick, J.J. Haller, Solenoid Actuator Having a Magnetic Flux Sensor, United States US5032812A, 1991, filed March 1, 1990, and issued July 16, 1991. <https://patents.google.com/patent/US5032812A/en>.
[5] K.W. Lim, N.C. Cheung, M.F. Rahman, Proportional control of a solenoid actuator, in: 20th International Conference on Industrial Electronics, Control, and Instrumentation, 1994, vol. 3. IECON '94, 1994, pp. 2045–2050. Available from: https://doi.org/10.1109/IECON.1994.398134.
[6] Y. Mitsutake, K. Hirata, Y. Ishihara, Dynamic response analysis of a linear solenoid actuator, IEEE Trans. Magn. 33 (2) (1997) 1634–1637. Available from: https://doi.org/10.1109/20.582582.
[7] M. Kamal, G.S. Daehn, A uniform pressure electromagnetic actuator for forming flat sheets, J. Manuf. Sci. Eng. 129 (2) (2007) 369–379. Available from: https://doi.org/10.1115/1.2515481.
[8] B.H. Shin, S.-W. Choi, Y.-B. Bang, S.-Y. Lee, An earthworm-like actuator using segmented solenoids, Smart Mater. Struct. 20 (10) (2011) 105020. Available from: https://doi.org/10.1088/0964-1726/20/10/105020.
[9] F. Fries, S. Miyashita, D. Rus, R. Pfeifer, D.D. Damian, Electromagnetically driven elastic actuator, in: 2014 IEEE International Conference on Robotics and Biomimetics (ROBIO 2014), 2014, pp. 309–314. Available from: https://doi.org/10.1109/ROBIO.2014.7090348.
[10] M.M. Said, J. Yunas, R.E. Pawinanto, B.Y. Majlis, B. Bais, PDMS based electromagnetic actuator membrane with embedded magnetic particles in polymer composite, Sens. Actuators A: Phys. 245 (July) (2016) 85–96. Available from: https://doi.org/10.1016/j.sna.2016.05.007.
[11] M. Rawlik, C. Crawford, A. Eggenberger, K. Kirch, J. Krempel, F.M. Piegsa, et al., A simple method of coil design. ArXiv:1709.04681 [Physics] 2017. <http://arxiv.org/abs/1709.04681>.
[12] R. Guo, L. Sheng, H.Y. Gong, J. Liu, Liquid metal spiral coil enabled soft electromagnetic actuator, Sci. China Technol. Sci. 61 (4) (2018) 516–521. Available from: https://doi.org/10.1007/s11431-017-9063-2.
[13] N. Ebrahimi, N.S. Gao, W. Taha Ahmad, J. Amir, Dynamic actuator selection and robust state-feedback control of networked soft actuators, in: Proceeding of IEEE International Conference on Robotic and Automation (ICRA 2018), 2018.
[14] W.K.H. Panofsky, M. Phillips, Classical Electricity and Magnetism, second ed., Courier Corporation, 2012.
[15] T.L. Pratt, Charge Model Expansion of the Standard Model of Particle Physics, BookSurge Publishing, 2008.
[16] E.P. Furlani, Permanent Magnet and Electromechanical Devices: Materials, Analysis, and Applications, Academic Press, 2001.
[17] W.T. Grandy Jr., Introduction to Electrodynamics and Radiation, Elsevier, 2012.
[18] D.C. Jiles, Introduction to Magnetism and Magnetic Materials, second ed., CRC Press, 1998.
[19] "American Wire Gauge Chart and AWG Electrical Current Load Limits Table with Skin Depth Frequencies and Wire Breaking Strength," <http://www.powerstream.com/Wire_Size.htm>, n.d. (accessed 26.07.18).
[20] P. Schimpf, A detailed explanation of solenoid force, Int. J. Recent. Trends Eng. Technol. (IJRTET) 8 (2013) 7–14.
[21] E.W. Weisstein, Circle Packing. <http://mathworld.wolfram.com/CirclePacking.html>, n.d. (accessed .8.05.18).
[22] "American Wire Gauge," Wikipedia. <https://en.wikipedia.org/w/index.php?title = American_wire_gauge&oldid = 849031042>, 2018.

[23] J.A. Miranda, Magnetic field calculation for arbitrarily shaped planar wires, Am. J. Phys. 68 (3) (2000) 254–258. Available from: https://doi.org/10.1119/1.19418.
[24] D.V. Schroeder, Entanglement isn't just for spin, Am. J. Phys. 85 (11) (2017) 812–820. Available from: https://doi.org/10.1119/1.5003808.
[25] C. Christodoulides, The magnetic field produced at a focus of a current-carrying conductor in the shape of a conic section, Am. J. Phys. 77 (2009) 1195–1196. Available from: https://doi.org/10.1119/1.3183888.
[26] W.R. Smythe, Static and Dynamic Electricity, first ed., CRC Press, New York, 1989.
[27] M.R. Spiegel, J. Liu, Mathematical Handbook of Formulas and Tables, McGraw-Hill, 1999.
[28] Mohamad Riahi, Nafiseh Ebrahimi, Test Apparatus for On-line Butt-Welding Evaluation of Aluminum Layer in PEX-AL-PEX Multilayer Pipes, Experimental Techniques 40 (2016) 185–193. Available from: https://doi.org/10.1007/s40799-016-0023-y.

CHAPTER 5

Cable-driven systems for robotic rehabilitation

Rand Hidayah, Tatiana Luna and Sunil Agrawal
Robotics and Rehabilitation Laboratory, Mechanical Engineering Department, Columbia University, New York, NY, United States

5.1 Introduction

Over the last several years, rehabilitation robotics has focused on how to best rehabilitate individuals with movement disorders such as stroke [1], spinal cord injury [2], or cerebral palsy [3]. Nearly 40 million adults in the United States have a mobility impairment which restricts their independence [4]. One of the strategies employed for restoring and rehabilitating movement with traditional physical therapy is repetitive training [5]. Repetitive training requires consistent and periodic motion applied to a patient's limbs or body in specific movement patterns. This particular type of rehabilitation is appropriate for a robotic device. Robotic devices can repeatedly and consistently provide forces to assist or perturb a user's motion. Furthermore, these devices can be engineered to apply controls and trajectories in a prescribed manner to suit physical therapists' needs for a specific patient [6,7]. The versatility of robotic controllers and architectures allow for varied strategies that can rehabilitate a range of human motion pathologies. This versatility is especially present in a specific architecture of rehabilitation robotic systems: cable-driven systems. Inherently, cable-driven systems do not add a lot of weight or inertia on the user. Cable-driven systems are also compliant, as they can address varied forms of human motion. This compliance is regarded in the sense of conforming to the specific kinematics and kinetics of the user in a particular motion. The systems can be made extremely lightweight and designed in ways to conform to both human anatomy and motion requirements. This allows for a single cable-driven system to adapt to several modes of human functional movement, and a simple rearrangement of the cable attachment and routing points be reconfigured to suit many different tasks.

Robotic devices for rehabilitation must consider factors such as safety, weight, cost-effectiveness, and capability to follow human motion. Rehabilitation robots, in general, are required to have high repeatability, reliability to follow trajectories, and the ability to apply forces to perform tasks with low risk when in close contact with human operators. Safety is critical as the robots are in contact with the human body. Rehabilitation robots must follow trajectories precisely and conform to the user's limb trajectories without interfering with the natural human motion. Forces applied must be precise and reliable, as effective assistance and training rely on correct support to the user within the rehabilitation context. For example, robots designed to support the body weight of the user, such as during complete spinal cord injury, must have high force and torque capability, when compared to robots which aim to retrain limb kinematics. Thus different rehabilitation scenarios may have different requirements for the robotic architecture.

There are two main robotic design choices that typically apply to rehabilitation applications: *rigid-link* and *cable-driven*. The Robotics and Rehabilitation Laboratory at Columbia University performs research with these devices for rehabilitation, but cable-driven devices are the focus of this chapter. *Rigid-link* exoskeletons have their appropriate applications, and will be introduced to provide context for their *cable-driven* counterparts. Rehabilitation robots typically act on the user at a single point of contact, as shown in Fig. 5.1, or could be worn on an entire limb, as in Fig. 5.2. These systems provide assistive forces for retraining functional movements [8] or bodyweight support to replace functional movements for users who cannot perform them independently [9].

Cable-driven rehabilitation robots also follow the same categorizations. There are devices that perform rehabilitative tasks by acting upon the end

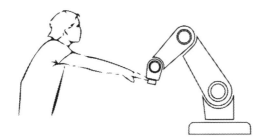

Figure 5.1 A point of contact end-effector-based system for the arm rehabilitation exoskeleton.

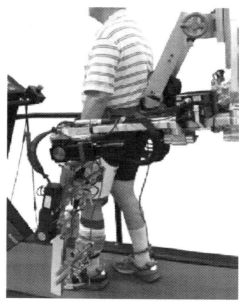

Figure 5.2 The ALEX II (Active Leg EXoskeleton), a worn exoskeleton developed by the Robotics and Rehabilitation Laboratory at Columbia University. The links provide torques and forces at the level of the entire limb.

effector, and devices which do so by acting upon the joints. End effector-based devices act on the human body and transfer trajectories, forces, or moments to a specific point, as shown in Fig. 5.3. An example of the other category of cable-driven rehabilitative device, an end-effector cable-driven device is shown in Fig. 5.4. The two architectures offer different biomechanical control strategies. The end-effector architecture offers the user more freedom on the configuration of their joints during a movement task, and typically uses end-point force field strategies to assist or inhibit a certain task. The worn exoskeleton architecture affects the end-effector position through the application of joint torques, which must follow the motions of the human limb rotation centers. Both architectures result in training gains but are appropriate in different tasks. The worn exoskeleton, for example, is appropriate in overground or mobile tasks, as well as tasks that require joint coordination and providing assistance through the joint torques.

Cable-driven systems provide many advantages compared to their rigid-link counterparts. The cables allow the off-boarding of motors, which allows less weight to be added to the user, and less inertia reflected

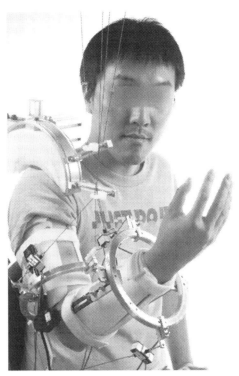

Figure 5.3 CAREX (Cable-driven Arm EXoskeleton): A cable-driven on user worn exoskeletal arm developed at the Robotics and Rehabilitation Laboratory at Columbia University providing end-effector trajectory and join torque assistance [10].

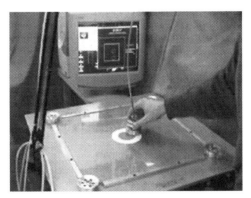

Figure 5.4 A point of contact planar table-top robot FeRiBa3 with a haptic display for arm trajectory tasks. The end-effector is connected to the motors through four cables [3]. *From P. Gallina, G. Rosati, A. Rossi, 3-d.o.f. wire driven planar haptic interface. J. Intell. Robot. Syst. 32 (2001) 23–36.*

on the user during movement [11]. Cable-driven systems allow motion of the human limbs with less obstruction [8]. Cables do not impose geometric constraints on the motion of the limbs, but provide torques and forces on the user to adapt their motion [12].

Cable-driven robots do possess certain disadvantages compared to their rigid-link counterparts. The estimation of the worn cable-driven robot is based on data-driven or anatomical models of the limbs, and thus always contains some error [13]. Friction, wear, and tear are design issues in most new prototypes, and methods to address these deficiencies are still currently under development to ensure the long life and robustness of the hardware, which their rigid-link counterparts do not typically consider [14,15]. Furthermore, in order to ensure safety and usability, appropriate cable materials [16] and nonslip features that attach to the human body must be selected to ensure that the system performs well and can be worn [17,18]. The stiffness of a cable-driven robot can vary based on the configuration of the human and the cable attachment points on the user, and is an important design selection when considering the goal and requirements of the user and the task which the system aims to assist [19].

5.1.1 Formulation of cable-driven devices

For cable-driven devices, an important feature is the *manipulability* of the system. Cables can only act in tension, or pull, this imposes an additional challenge on the design of the system. It has been shown that cable-driven systems of n degrees-of-freedom require a minimum of $n + 1$ cable actuators to be completely restrained [20−23].

The general form of the force equilibrium equations for a fully restrained cable-driven system is as follows:

$$F = J^T T \tag{5.1}$$

$$F = \begin{bmatrix} F \\ \vdots \\ M \end{bmatrix}, \quad J^T = \begin{bmatrix} u_{1,1} & & u_{n,1} \\ \vdots & & \vdots \\ u_{n+1} & & u_{n+1,n} \end{bmatrix}, \quad T = \begin{bmatrix} T_1 \\ \vdots \\ T_{n+1} \end{bmatrix} \tag{5.2}$$

where F is the generalized forces acting upon the point of interest in the system, J^T is the transpose of the Jacobian of the manipulator, and T is the tensions of the cables which are acting upon the end effector of a serial chain. As mentioned, $n + 1$ cables are required to fully restrain the system. However, if more than $n + 1$ cables actuate the system, then the system is redundant but still manipulable. Reasons to expand a cable-driven system to a redundant system include workspace requirements, power-dissipation or force ceiling impositions, and cable positioning requirements for specific pose restraint or for systems under specific loads [20,24–29].

5.1.1.1 Device work-space

For a cable-driven device, the space in which it is able to apply forces through its end-effector is described as its wrench feasible workspace [30]. This is a subset of the larger workspace which is characterized by all possible combinations of cable lengths pertaining to a specific cable attachment point arrangement and a cable origin arrangement, such as in Fig. 5.5.

This is of particular importance for cable-driven rehabilitative systems, where the capabilities and motion of the user need to be assessed and used as a design requirement [31]. If the wrench feasible workspace does not match with the workspace of the human limb, say in a walking or reaching task, then the appropriate forces cannot be applied to the user's limb for assistive forces. This results in cable-driven systems losing tension (becoming slack) and the failure of the system. Some forms of failure recovery are available for cable-driven systems [32,33]. One such recovery tool is maintaining a minimal tension and detecting end-effector positions which are outside of the workspace.

Within a wrench feasible workspace, there are specific poses and positions in which the cable-driven robot is more *dexterous* [25,34]. These dexterity indices describe where the cable-driven system can more easily

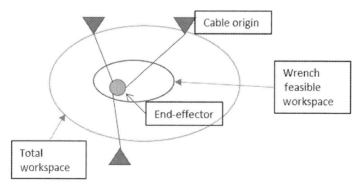

Figure 5.5 A representation of the workspace available for all cable lengths and positions and the wrench feasible workspace of a system which allows the required forces to be applied to the end effector.

change position, orientation, and velocity of the end effector. These indices are used to evaluate different *cable attachment points* and *cable routing points* and to compare between designs of a cable-driven device which requires the same degrees-of-freedom at the end effector, a similar wrench, and the same number of actuators or cables in order to achieve the required task [35,36].

5.1.1.2 Control schema considerations: toward compliant controls

Cable-driven systems have control schemas of various types in order to follow or apply forces on the limb of a user for a rehabilitative task. In terms of sensor feedback, cable-driven systems have the option of using cable-length information via motor encoders [37], or cable tensions to apply closed-loop tension control [10], or by assessing the end effector's trajectory through motion tracking sensors [38].

Cable-driven systems can be made reasonably transparent to the user due to low inertia and weight of the cables [15]. The stiffness of the controller allows compliance to the user's motion and different magnitudes of assistance can yield different biomechanical results [37]. Typical human biomechanics has been characterized for several tasks and for normative populations [39]. How humans learn new movement tasks has also been studied and characterized [40]. Roboticists have implemented control strategies that reflect assistive training [41], as well as resistive training [42], in order to encourage neuromuscular activity and improve the system's proficiency in performing a movement. This encourages the rehabilitation of motion and achieving better quality movements [43,44].

The strategies used to characterize the effect a robotic device on a user may vary. Some devices sense internal forces and torques which are the operational bases of the system [45]. Muscle activation of the user is used as a metric for the robot assistance [46–48]. The metabolic cost of performing a task is also used as a measure to augment human capabilities with a worn exoskeleton [37,49,50]. There is a trend toward providing strategic assistance to a patient through the control scheme which is based on compliance to human movement, as opposed to a prescriptive trajectory [8,51,52]. This follows motor learning principles on performance and retention of motion, where it has been shown that variable and challenging strategies result in a better retention of a new learning movement pattern [53] and a better outcome for a patient population [43,54]. Learning how the user is adapting to the intervention is also a development in control algorithms which helps modulate the therapy over time [55].

5.1.2 Categories of cable-driven devices

We present some practical examples of cable-driven systems that are end-effector based, worn on the user's body, and are used for specific rehabilitation purposes. These devices may have one of two cable attachment configurations: parallel or serial.

5.1.2.1 Serial cable-driven devices

Serial cable-driven exoskeletons follow the serial skeletal structure of a human limb. The cables pull at links that are analogous to the human skeletal structure, such as in the gait training robot LOPES [56]. These serial cable-driven devices can have a soft structure, such as finger-based tendon drive exoskeletons [57]. While these cable-driven devices depend on cables for actuation, they typically impose a geometric constraint as each part of the exoskeleton is still connected through mechanical joints. They still require $n + 1$ cables for fully restrained actuation, but can precisely formulate a model from the geometric and mass properties similar to rigid-link exoskeletons [13].

5.1.2.2 Parallel cable-driven devices

Parallel cable-driven devices are the focus of many rehabilitation efforts of the lower limbs and torso [58,59]. The parallel nature of the cables and their attachments to the human through a brace, cuff, or orthosis requires the assistive strategies be based on applying joint-level forces or torques. The user is encouraged to adapt their end-effector trajectory, which is manipulated by coordinated forces applied to the joints [60]. The user's body can also be given a set-point force in order to explore the movement patterns which result from a specific motion task with an external robotic force application [58].

5.1.3 Cable-driven systems for postural and gait rehabilitation with variable controllers

Our work is primarily focused on parallel mechanisms that offer compliance to human motion thought control schemes which assist, support, and challenge the user according to their own motion and mass properties [61–64]. This allows the robotic exoskeletons to induce motion adaptation through varied and proportionate assistance. This addresses the problem of overdependence on feedback, which can inhibit efficient motor learning by not being compliant to user movements and restricting movement variability in order to ascertain efficient movement patterns [40].

5.1.3.1 Cable-driven Active Leg Exoskeleton

The Cable-driven Active Leg Exoskeleton (C-ALEX) is a parallel, worn exoskeleton which applies joint torques through tensions in the cables. These joints torques correspond to a desired force profile at the ankle level and direct a desired gait trajectory [60]. The controller is formulated as follows and the biomechanics are shown in Fig. 5.6.

The distance d of P_{ankle} with respect to P' dictates the amount of normal and tangential force assistance that is required at the ankle level. This is defined as:

$$F_a = \begin{bmatrix} K_n(1 - e^{-\lambda d^2}) \\ K_t e^{-\lambda d^2} \end{bmatrix} \quad (5.3)$$

where K_n and K_t are gain constants set by the controller to prescribe the assistance for that session. λ is a scaling factor that expands or retracts the tunnel around the gait trajectory where the force field begins to act. These forces are related to assistive joint torques by the serial human limb Jacobian:

$$\tau_a = J_E F_a \quad (5.4)$$

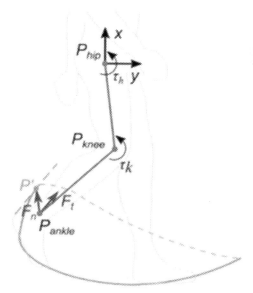

Figure 5.6 Force assistance as determined by the target gait trajectory of the C-ALEX system.

This assistive torque is then provided by the cable tensions of the exoskeletal system, and are based on the cable Jacobian.

$$\tau_a = J_T T \qquad (5.5)$$

The system is also subject to a minimum tension as well as a quadratic solver to find a continuous and smooth solution of the system [60]. This schema provides an assist-as-needed control paradigm that depends on the control gains and the force tunnel around the target gait trajectory. If a user is following the trajectory within a prescribed tolerance, then the controller provides no or little assistance. This approach is followed to allow a user to learn by following the force interaction and try different strategies to reach a desirable gait or joint trajectory while wearing the exoskeleton. The approach is chosen to remain compliant to the user's performance in following the prescribed gait. This approach is motivated by the best practices of optimal motor learning [65]. Case studies with the C-ALEX system showcasing the effects of this assist-as-needed strategy in an overground task and on treadmill training are discussed in Section 5.2.

The control of C-ALEX depends on the high-level mathematical formulation extending current commands to a motor. Based on the cable tension planner which dictates the cable tensions in Eq. 5.5, the proportional integral derivative controller of the C-ALEX system attempts to match the current commands at 1000 Hz. This controller schema is always based on the target path. The controller block diagram is shown in Fig. 5.7.

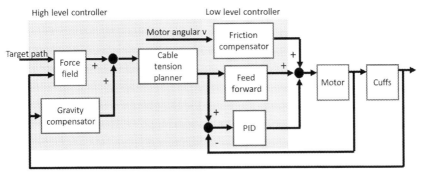

Figure 5.7 Controller block diagram of the C-ALEX system showing the high-level controller which formulates the overall target trajectory goals and the low-level controller which commands the motor currents and attempts to match the desired cable tensions.

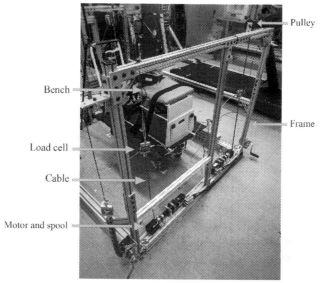

Figure 5.8 TruST (Trunk Support Trainer) with its components, the 80/20 frame, the cables are routed from the belt to load cell to the motor. TruST in this configuration would be used to train a subject seated on the bench for posture rehabilitation.

5.1.3.2 The Trunk Support Trainer

The Trunk Support Trainer (TruST) is another cable-driven robotic platform used for posture training and rehabilitation [66]. This robotic trainer uses a planar configuration to control the movement of a user's trunk by connecting four cables to a belt. Each cable is then routed to the end of a load cell and then connected to the end of a Direct Current motor, Fig. 5.8. The load cells provide the current tension values in real time.

TruST can be used in the following modes: (1) transparent mode, a minimum tension is maintained throughout the subject's movements; (2) perturbation mode, a force is applied in one of eight possible directions, anterior, posterior, left or right lateral sides, and the four diagonal directions, within the plane. This force can be modified by the user to adjust the magnitude, direction, or duration; (3) constant mode, a set tension value is maintained for all four cables; and (4) force field mode, a boundary is defined where TruST is in transparent mode and once the subject is outside of this defined boundary a force is applied in the direction of the subject's center. The magnitude of the force is chosen to be a bodyweight percentage. Reflective markers are placed on the belt, and with the use of a motion capture system, the position of the belt and thus the cable endpoints are determined.

There have been several studies performed utilizing TruST's control modes. The first study evaluated the effects TruST had on the kinematics and center of mass (COM) of 10 able-bodied subjects when seated. They obtained the assistance of the force field and were compared to a control group of 10 other subjects that received no assistance. This was a proof of concept study that showed that the use of training with TruST can increase a subject's seated workspace after one training session [66].

Another study was conducted to compare the effects of virtual reality on posture training with and without the use of TruST's force field assist-as-needed mode [67]. The results showed that the use of virtual reality and TruST had similar effects to the use of TruST alone, but both increased the subjects' kinematic COM range compared to the subjects that did not receive training with the assistance of TruST. These previous experiments were performed on able-bodied subjects to show the potential of training with TruST to increase seated reaching workspace. This was used to validate the use of training with other populations that lack postural control.

The TruST has been modified to also assist in postural training while standing. This is achieved by attaching a belt to the trunk and to the pelvis of a subject. The trunk is actuated by the TruST motors. Meanwhile, the pelvis belt is connected to cables with springs that provide a passive resistance. A pilot study was performed to show that training in this configuration improves subjects' upper body postural stability [64]. This study contributed to the development of another robotic device that focuses on posture training solely while standing, the Robotic Upright Stand Trainer (RobUST).

5.1.3.3 Robotic Upright Stand Trainer

Another cable-driven device that has been developed to train posture is the RobUST [63]. RobUST has 14 motors to actuate the trunk, pelvis, and knees. There are four cables routed to the trunk belt, up to eight cables routed to the pelvis belt, and one cable routed to each knee. Each cable is actuated by an individual motor and is passed through a load cell to determine the tension value (Fig. 5.9). The position of each belt is tracked by a motion capture system.

RobUST has a low-level and high-level controller. The low-level controller determines what current command to send to the motors to

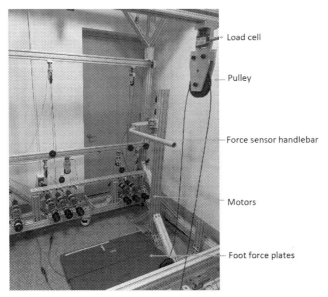

Figure 5.9 Image of RobUST with two force plates and a force sensor handlebar in the center of the frame.

achieve the desired tension. The high-level controller determines the desired forces and calculates the desired tensions values. These forces depend on the control mode selected from the user interface.

The trunk and pelvis can be controlled in any of the following modes: (1) transparent mode; (2) constant mode; (3) perturbation mode; and (4) force field mode. The transparent mode keeps a minimal tension value of 18 N per cable. The constant mode maintains a user-defined constant tension value per cable. The perturbation mode is similar to the TruST's perturbation mode, it provides a force that can be modified by the user.

The force field mode has different variations but all utilize an assist-as-needed approach. When a subject is within a defined boundary, the system is in transparent mode. Once outside this boundary, the subject is assisted by RobUST. The boundary and the assistance force varies with the type of force field. The planar force fields only solve for desired forces in a single plane. In planar mode, only four cables are needed at the trunk and four at the pelvis. RobUST is also able to achieve a 3D force field at the pelvis, which is when eight cables are connected to the pelvic belt. This 3D force field solves for desired forces in the three directions, x, y, and z. The boundary of the force field can vary, it can be subject dependent or can have a uniform area. In the study by Khan et al. [63], the force field boundary was defined as a circle with a constant radius for all subjects.

RobUST was developed to train postural stability, Section 5.3 covers further information of a case study utilizing RobUST in perturbation and force field modes with a specific use.

5.1.3.4 Tethered Pelvic Assist Device

The Tethered Pelvic Assist Device or TPAD is a cable robot that applies forces to the human pelvis and causes a gait adaptation [58]. A user in the system is shown in Fig. 5.10A. TPAD can both impose a planar control or a spatial control depending on the desired gait adaptation. Both modalities of planar and spatial trajectory and force applications have the same controller shown in Fig. 5.10B. The TPAD is a very low inertia device that aids in creating force fields and assistive control strategies that add minimal dynamics to the user. The forces are provided almost exclusively and in many different directions. These include studies on crouch gait in cerebral palsy children [61], inducing asymmetry in adults [68,69] perturbation training in healthy [70], stroke asymmetry [71,72], and ataxia [73].

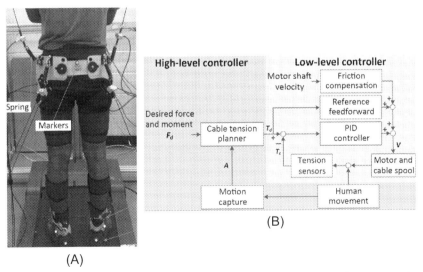

Figure 5.10 Overview of the active Tethered Pelvic Assist Device. (A) A user in the TPAD system walking on a treadmill, markers and springs attached to the system which are used in the control methodology are shown. (B) Controller for spatial and planar applications of forces through the ATPAD system.

TPAD also uses an underdetermined tension solver with a quadratic program to solve the tensions for desired forces and moments applied to the human body [74]. The forces can be simple or complex as required by the task. A single upward or downward force with no targeted moments can be applied at specific times in the gait cycle [71]. Similarly, targeted lateral forces to change trajectories can be applied with the same controller, only changing the positional desired path parameters [69]. The TPAD thus similarly complies to the user requirement of training for specific tasks.

5.2 Cable-driven leg exoskeleton for gait rehabilitation

C-ALEX, shown in Fig. 5.11, uses an assist-as-needed control schema in two case studies. One includes the exploration of the effect on pressure distributions of hemiplegic poststroke individuals. The second involves a step-up task with healthy users in an augmented reality environment.

A common issue with leg exoskeletons is the joint level center of rotation misalignment between the robot joint centers and anatomical joint centers [11,13]. C-ALEX does not inhibit motion or the natural degrees-of-freedom of the human limb. Thus it can be used in more functional movements requiring abduction and rotation [15]. The system also does not restrict pathological movements present after stroke events [1] while still providing repeatable and consistent feedback to a walking task. This approach to rehabilitating movement is of interest for early stroke recovery [5].

Differences in gait overground and on a treadmill also require that exoskeletons be versatile to both [75]. The C-ALEX architecture can be adapted to a mobile platform. Mobile platforms must deal with various forms of movement and terrain. This may require different kinematics than the highly regular leg movement during walking on a treadmill, that many leg exoskeletons base their controllers on. C-ALEX has been shown to have similar performance overground and on a treadmill [62]. The versatility of this system lends itself to augmenting both limb trajectories and joint angle profiles, depending on the challenge presented in the task.

5.2.1 Stroke rehabilitation case study

A single-session training of poststroke individuals was carried out with the C-ALEX system providing assistive forces in order to assess the effects of increased step length and step height on the pressure distribution and

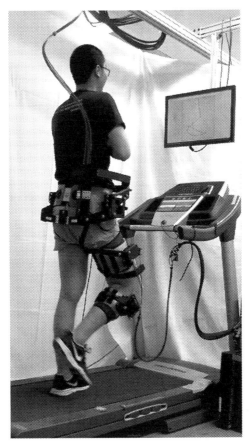

Figure 5.11 The Cable-driven Active Leg Exoskeleton, or C-ALEX system, used on a participant in a gait training protocol. The system is based on a cable tension planner that provides joint torques at the knee and hip level to assist the ankle end-effector to follow a trajectory. This trajectory is meant to actively cause an adaptation within a training task. In this case, walking.

footfall characteristics of the user. User behavior in redistributing their weight while walking is a key question of stroke rehabilitation [71]. The results from this study are based on the protocol outlined in Fig. 5.12. The C-ALEX system was fitted to the participants and familiarization with walking with the system was carried out. This was done in order to ascertain the comfortable walking speeds and for the physical therapist to decide upon the magnitude of the increase in each step parameter. After this, the effects of wearing the exoskeleton were characterized in a training protocol targeting specific length and height increases [76].

Cable-driven systems for robotic rehabilitation   151

Figure 5.12 Protocol for training in a single-session of poststroke individuals. Data retained was kinematics and foot pressure on the instrumented treadmill.

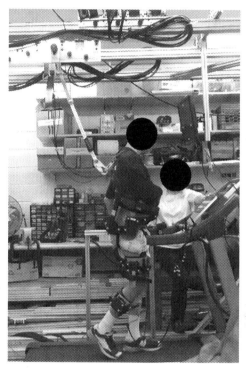

Figure 5.13 The C-ALEX system, used with a poststroke participant in a gait training protocol. A physical therapist observes and dictates the training goals in the protocol. The goal of this training was to explore the effects of increasing step length and step height. The effects of the system on symmetry and pressure line data are presented in this section.

C-ALEX on a participant with the physical therapist observing the changes in gait on the side is shown in Fig. 5.13. The C-ALEX system has the unique capability of showing real-time feedback to the user and to the physical therapist. While this is not part of this protocol, this allows both users of the system to use different training strategies, either by assisting the users with larger forces that support motion more strongly or by resisting movements or perturbing them to increase the variability

COP diagram

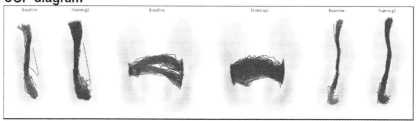

Figure 5.14 The variation of center of pressure parameters between a training session and a baseline session. In this session, the user expands the area of the foot where the line of gait is applied on the paretic (left) side of the body. The overall center of pressure butterfly diagram in the middle, which shows the center of pressure over many gait cycles, is actually more symmetrical in where the loading happens during training thin baseline. The blue lines cross over more of the paretic side's foot during the training sessions than during baseline sessions.

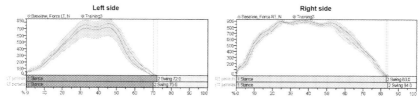

Figure 5.15 Force loading diagrams of a single subject during training and baseline. The highly varied force loading pattern on the paretic (left) side shows that the subject is exploring a different adaptation strategy when compared to the baseline measure [40]. The optimal force loading pattern would ba symmetric pattern on both legs, which would be the goal of multiple sessions of gait retraining and what robotic and modern gait therapies target after early stroke events [5].

surrounding the environment of task [77]. The kinematics of the paretic limb were targeted, and the overall effect on the gait line and center of pressure (COP) was considered.

A representative participant's results are shown in Figs. 5.14 and 5.15. The results of this case study focused on the foot loading characteristics of the participants as a group, but the results of a single representative subject is shown in both figures. The effect of changing and adapting kinematics on the loading and gait line characteristics of both legs is evaluated, motivated by the need to characterize the effects of an intervention on gait symmetry poststroke.

The results from one representative subject show the change in step length and step height in a new loading pattern. There is an observed increase in the loading envelopes of the paretic leg. The effective area of the COP of the foot is expanded. The gait line of the foot on the paretic side is longer. The increase in the area suggests that the stable base of support has changed. This could indicate that the user is exploring other strategies to achieve a specific kinematic target and the kinematic target requires a different loading pattern. This is of interest to a physical therapist, as symmetry in loading and symmetry in the COP distribution while walking makes an inherently unstable task more dynamically stable. This link between kinematics directly affecting the loading patterns encourages interventions at the joint level for affecting symmetry.

Overall, the participants showed an adaptation of gait parameters. We consider three sessions for the discussion of the results:

1. Pretraining (PRT), which is the sham exoskeleton worn, compensating for its own self-weight but not providing assistance.
2. Early training session (T1), which is the first bout of training which the participant undertakes.
3. Late training session (T4), which is the last bout of training which the participant undertakes.

The chosen sessions show the effect of the robot without providing assistance, which is important to characterize as it shows the system's effectiveness with regard to baseline. The early training session was selected to showcase the early adaptation results of being in the system. The late training session results were selected to showcase the effect of the system after a familiarization period of nearly 30 min. These results provide a snapshot of the effects of the system at various periods of familiarity and challenge to the user. For the length of the gait line difference, there is a decrease across all subjects in this protocol for the training sessions. The effect of the exoskeleton is to decrease the differences in gait line symmetry. Gait line symmetry denotes the length of the entire gait line—the butterfly graph in Fig. 5.14. The length of the line from left to right and the converse are compared. The differences provide a percentage difference between the paretic and nonparetic leg. The values of this gait line difference are then normalized by the baseline session value. This gives insight into the magnitudes of the decrease across the training sessions. The subject by subject data is presented to denote the differences between users inherent in the system.

The asymmetry of loading of the paretic and nonparetic leg is also considered in Fig. 5.16B. Again, this value is compared to the subjects' baseline, as this is the metric that best shows what the effect of the C-

154 Soft Robotics in Rehabilitation

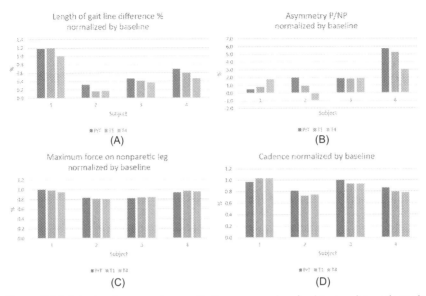

Figure 5.16 Select gait parameter results from a sample of subjects who undertook the training outlined in Fig. 5.12. The subjects showed a gait adaptation of loading parameters during the training sessions. The effect of changing overall kinematics of the leg during gait, on one side, is seen to potentially change the loading parameters which pertain to overall symmetry. (A) Normalized length of gait line difference results for four stroke participants in the protocol. (B) Asymmetry between the paretic and nonparetic side of the participants gait during training. (C) Maximum force normalized by baseline of the nonparetic leg during training sessions. (D) Cadence of the steps normalized by the baseline of the subjects during training.

ALEX system is while being worn. This was the most inconsistent result based on the comparison to baseline. Certain subjects increased the asymmetry, others decreased or flipped the asymmetry direction to the paretic side. While a trend was observed here, it is interesting to consider that overall effects of asymmetry are not conclusive through changes in kinematics, though there are trends in other loading parameters and gait line formation.

The force on the nonparetic leg stays nearly constant but is overall decreased from baseline as demonstrated in Fig. 5.16C. This is an interesting result when combined with the length of the gait line. This analysis suggests that the added length on the paretic leg means that the double stance phase is where the overall decrease in max force occurs. With a max force decrease and a decrease in the difference between the gait line of the paretic and nonparetic side, this suggests that a longer paretic gait line is being loaded throughout all subjects in the training protocol.

The cadence is either constant or decreased by nearly 20% as shown in Fig. 5.16D. This is an overall speed metric that aims to show whether shorter steps are being taken in a shuffling motion at a set speed. The decrease is encouraging, as this suggests that certain users gained some confidence in changing their step number for a set speed. The slower steps are likely more energetically appropriate [39] and suggest an adaptation that can be focused on for more practical gait retraining purposes in a clinical setting or protocol.

5.2.2 Augmented Reality stepping task study

The use of augmented reality (AR) with overground exoskeleton tasks is an understudied field of research [59]. Variability in feedback is a vital part of motor learning [40], and the lack of options for overground tasks in terms of visual representation of the user is a clear deficiency [15]. The C-ALEX overground module, shown in Fig. 5.17A, was developed in order to explore overground functional tasks and the effect of the C-ALEX system can be studied both using visual and haptic feedback. In order to assess nonwalking tasks in a variable environment, a stepping-up task was explored on a group of healthy users.

The stepping stones task required the user to wear C-ALEX and to clear a series of step ups randomly placed in their footpath. A representation of a partial footpath with the stepping holograms is shown in Fig. 5.17B. As the user steps onto the path, she must clear her foot while looking down at the obstacle which appears. The C-ALEX system tracks and compensates for its own gravity while this experiment was performed. Preliminary results show the effect of the augmented reality on the perception and action of the user, as her user behavior is changed from one step height to the next.

The results are shown in Fig. 5.18 where the user behavior changes when coming into contact with a perceived new obstacle height. The data shown is of one user in the system, illustrating the user's change in trajectory when navigating over steps increasing in height. While the steps were placed randomly, the results are ordered from lowest to highest in the figure legend to illustrate the variation.

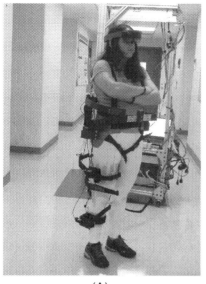
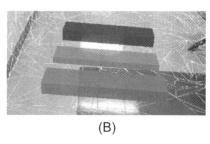

Figure 5.17 User and holograms in the augmented reality environment during the stepping-up task. The C-ALEX system compensates for its own gravity as the user attempts to navigate and clear obstacles on the ground in the augmented reality environment. The obstacles are superimposed over the uninhibited view of the user. (A) The mobile module of the C-ALEX system with a user wearing both the leg for haptic feedback and the HoloLens for visual feedback during a mobile overground task. In this case, a step-over task is targeted repetitively while increasing the step block heights. (B) The mobile module of the C-ALEX system with a user wearing both the leg for haptic feedback and the HoloLens for visual feedback during a mobile overground task. In this case, a step-over task is targeted repetitively while increasing the step block heights.

5.3 A perturbation study using Robotic Upright Stand Trainer

A single-subject case study was performed using the RobUST. The aim of the study was to set the groundwork for a further ten subject study [63]. Perturbations were applied randomly to a user in four directions, twice. The application was done at the trunk and pelvis level. RobUST would provide a perturbing force at one level and either an assisting force field or no assistance at the other level. This was then repeated with the levels switched.

The motivation for this case study was to determine the magnitude and force profile to create instability. This instability would be defined as

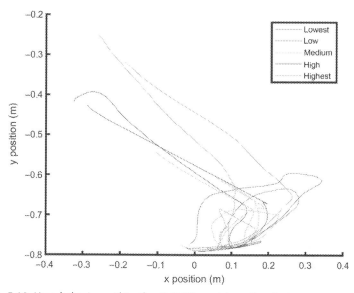

Figure 5.18 User behavior within the C-ALEX module acting in transparent mode or compensating for its own self-weight while performing a step-up task in an augmented reality environment. These preliminary results show a very different user behavior when perceiving the different obstacles in AR, and are encouraging for the exploratory direction for robotics, human-in-the-loop interaction, and augmented reality environments. The advantages of using AR with rehabilitation is the ability to vary the environments and types of referential feedback without imposing physical obstacles which can be risk factors in terms of tripping or collisions with the human limbs.

a perturbation force. This was repeatedly tested at the trunk and pelvis level. The effects of the subject's COM, was recorded using a motion capture system. Another motivation for this study was to also characterize the effect that the RobUST assist-as-needed force field had on the COM during the different perturbation levels. One can see with the use of the force field the displacement of the COM is less in all scenarios, Fig. 5.19.

To further challenge postural balance and stability, the perturbations were repeated while the subject was standing on a hemisphere balance ball. In this setup, the perturbations were only provided at the trunk level. The motivation for this was that perturbation at the trunk level create the greatest trunk instability, as seen in [63]. Fig. 5.20 depicts an example of the COP, recorded from the force plate during a perturbation. The COP represents the hemisphere balance ball's application of force placed on top of a force plate, and imposes a limitation of interpreting the results in terms of user behavior, when contrasted with direct foot contact with a

158 Soft Robotics in Rehabilitation

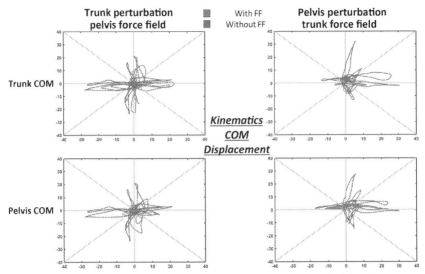

Figure 5.19 A subject's center of mass displacement, as recorded by the motion capture system, during perturbation forces in the anterior, posterior, and lateral sides.

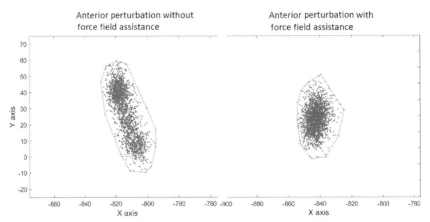

Figure 5.20 While the subject was standing on a balance ball placed on top of a force plate, the center of pressure of the ball was recorded. This is during an anterior perturbation at the trunk.

force plate. However, with the force field, there was less displacement of the COP, which is interesting in a system which requires balance compensation from the user due to a more challenging environment.

5.4 Conclusion

Wearable Cable-driven systems are versatile and well-positioned for use with rehabilitative robotics. Parallel cable-driven architectures are particularly well-suited for these tasks, as they do not impose any constraints on the geometry of the user and can be adjusted to different limb and body types.

These cable-driven systems face challenges in robust sensing and control. However they are more advantageous as they are more transparent to the user, and by being nonrestrictive and adaptable to different body types. The trade-off requires better implementations of control that provide feedback based on a model or data-driven strategies, as well as the exploration of different materials for the cables that actuate the user's limbs, as well as the human-robot interface in the donnable modules.

The compliance in cable-driven systems lies within their controllers. They are able to more transparently apply force and trajectory strategies to a user. Such strategies range from applying set forces and torques to assistive or resistive force profiles. With these strategies, the cable-driven control systems can range from very stiff to very compliant. They are unique in their ability to be fully ergonomic to the user, without compromising the reliability of a traditional rigid-link robotic system.

Acknowledgments

We would like to thank Siddharth Chamarthy, Dr. Lauri Bishop, Dr. Xin Jin, and Dr. Moiz Khan for their assistance with the C-ALEX and RoBUST devices.

References

[1] B.H. Dobkin, Strategies for stroke rehabilitation, Lancet Neurol. 3 (9) (2004) 528–536.
[2] T. Lam, J.J. Eng, D.L. Wolfe, J.T. Hsieh, M. Whittaker, Scire Research Team, A systematic review of the efficacy of gait rehabilitation strategies for spinal cord injury, Top. Spinal Cord. Inj. Rehabil. 13 (1) (2007) 32–57.
[3] D.L. Damiano, Rehabilitative therapies in cerebral palsy: the good, the not as good, and the possible NIH public access, J. Child. Neurol. 24 (9) (2009) 1200–1204.
[4] Centers for Disease Control and Prevention, Difficulties in physical functioning among adults aged 18 and over, by selected characteristics: United States, 2014. Technical report, National Center for Health Statistics, 2014.
[5] J. Schröder, S. Truijen, T. Van Criekinge, W. Saeys, Feasibility and effectiveness of repetitive gait training early after stroke: a systematic review and meta-analysis, J. Rehabil. Med. 51 (2) (2019) 78–88.

[6] C. Bayon, R. Raya, S.L. Lara, O. Ramirez, I. Serrano J, E. Rocon, Robotic therapies for children with cerebral palsy: a systematic review, Transl. Biomed. 7 (1) (2016).
[7] M. Babaiasl, S.H. Mahdioun, P. Jaryani, M. Yazdani, A review of technological and clinical aspects of robot-aided rehabilitation of upper-extremity after stroke, Disabil. Rehabil.: Assist. Technol. 11 (4) (2015) 1−18.
[8] A.S. Gorgey, Robotic exoskeletons: the current pros and cons, World J. Orthop. 9 (9) (2018) 112−119.
[9] M. Chiara, C. Silvestro, M. Jos, International symposium, and wearable robotics, *Wearable Robotics: Challenges and Trends*, vol. 16 of Biosystems & Biorobotics, Springer International Publishing, Cham, 2017.
[10] Y. Mao, S.K. Agrawal, Design of a cable-driven arm exoskeleton (CAREX) for neural rehabilitation, IEEE Trans. Robot. 28 (4) (2012) 922−931.
[11] D. Zanotto, Y. Akiyama, P. Stegall, S.K. Agrawal, Knee joint misalignment in exoskeletons for the lower extremities: effects on user's gait, IEEE Trans. Robot. 31 (4) (2015) 978−987.
[12] R. Gassert, V. Dietz, Rehabilitation robots for the treatment of sensorimotor deficits: a neurophysiological perspective, J. Neuroeng. Rehabil. 15 (1) (2018) 46.
[13] H. Xiong, X. Diao, A review of cable-driven rehabilitation devices, Disabil. Rehabil.: Assist. Technol. (2019) 1−13.
[14] J.-Y. Kuan, K.A. Pasch, H.M. Herr, A high-performance cable-drive module for the development of wearable devices, IEEE/ASME Trans. Mechatron. 23 (3) (2018) 1238−1248.
[15] A.J. Young, D.P. Ferris, State of the art and future directions for lower limb robotic exoskeletons, IEEE Trans. Neural Syst. Rehabil. Eng. 25 (2) (2017) 171−182.
[16] A. Mazumdar, S.J. Spencer, C. Hobart, J. Dabling, T. Blada, K. Dullea, et al., Synthetic fiber capstan drives for highly efficient, torque controlled, robotic applications, IEEE Robot. Autom. Lett. 2 (2) (2017) 554−561.
[17] A.T. Asbeck, R.J. Dyer, A.F. Larusson, C.J. Walsh, Biologically-inspired soft exosuit, in: IEEE International Conference on Rehabilitation Robotics, 2013, pp. 1−8.
[18] F. Connolly, D.A. Wagner, J. Walsh, K. Bertoldi, J.A. Paulson, Sew-free anisotropic textile composites for rapid design and manufacturing of soft wearable robots, Extreme Mech. Lett. 27 (2019) 52−58.
[19] J. Piao, J. Jung, J.O. Park, S.Y. Ko, S. Park, Analysis of configuration of planar cable-driven parallel robot on natural frequency, in: 2016 IEEE International Conference on Robotics and Biomimetics, ROBIO 2016, 2016, pp. 1588−1593.
[20] S.K. Mustafa, S.K. Agrawal, Force-closure of spring-loaded cable-driven open chains: minimum number of cables required & influence of spring placements, in: 2012 IEEE International Conference on Robotics and Automation, IEEE, 2012, pp. 1482−1487.
[21] R.L. Williams, P. Gallina, Planar cable-direct-driven robots: design for wrench exertion, J. Intell. Robotic Systems: Theory Appl. 35 (2) (2002) 203−219.
[22] R.L. Williams, P. Gallina, Planar cable-direct-driven robots, part I: Kinematics and statics, in: Proceedings of the ASME Design Engineering Technical Conference, 2001.
[23] P. Gallina, A. Rossi, R.L. Williams, Planar cable-direct-driven robots, Part II: Dynamics and control, in: Proceedings of the ASME Design Engineering Technical Conference, 2001.
[24] J.T. Bryson, X. Jin, S.K. Agrawal, Optimal design of cable-driven manipulators using particle swarm optimization, J. Mech. Robot. 8 (4) (2016).
[25] G. Rosati, D. Zanotto, A novel perspective in the design of cable-driven systems, in: ASME International Mechanical Engineering Congress and Exposition, Proceedings, 2009.

[26] L. Notash, Antipodal criteria for workspace characterization of spatial cable-driven robots, in: A. Pott, T. Bruckmann (Eds.), Cable-Driven Parallel Robots. CableCon, 2019. Mechanisms and Machine Science, vol 74. Springer, Cham, 2019. https://doi.org/10.1007/978-3-030-20751-9_17.
[27] E. Stump, V. Kumar, Workspaces of cable-actuated parallel manipulators, J. Mech. Design Trans. ASME 128 (1) (2006) 159—167.
[28] H. Wang, J. Kinugawa, K. Kosuge, Exact kinematic modeling and identification of reconfigurable cable-driven robots with dual-pulley cable guiding mechanisms, IEEE/ASME Trans. Mechatron. 24 (2) (2019) 774—784.
[29] Y. Wang, C. Song, T. Zheng, D. Lau, K. Yang, G. Yang, Cable routing design and performance evaluation for multi-link cable-driven robots with minimal number of actuating cables, IEEE Access. 7 (2019) 135790—135800.
[30] P. Bosscher, I. Ebert-Uphoff, Wrench-based analysis of cable-driven robots, in: IEEE International Conference on Robotics and Automation, 2004. Proceedings. ICRA '04. 2004, IEEE, April 2004, pp. 4950—4955.
[31] S. Bouchard, C. Gosselin, B. Moore, On the ability of a cable-driven robot to generate a prescribed set of wrenches, J. Mech. Robot. 2 (1) (2010) 1—10.
[32] L. Notash, Failure recovery for wrench capability of wire-actuated parallel manipulators, Robotica 30 (6) (2012) 941—950.
[33] G. Boschetti, G. Carbone, C. Passarini, Cable failure operation strategy for a rehabilitation cable-driven robot, Robotics 8 (1) (2019) 17.
[34] L. Wu, R. Crawford, J. Roberts, Dexterity analysis of three 6-DOF continuum robots combining concentric tube mechanisms and cable-driven mechanisms, IEEE Robot. Autom. Lett. (2017).
[35] D. Lau, D. Oetomo, S.K. Halgamuge, Generalized modeling of multilink cable-driven manipulators with arbitrary routing using the cable-routing matrix, IEEE Trans. Robot. 29 (5) (2013) 1102—1113.
[36] G. Rosati, D. Zanotto, R. Secoli, A. Rossi, Design and control of two planar cable-driven robots for upper-limb neurorehabilitation, in: 2009 IEEE International Conference on Rehabilitation Robotics, ICORR 2009, 2009, pp. 560—565.
[37] B.T. Quinlivan, S. Lee, P. Malcolm, D.M. Rossi, M. Grimmer, C. Siviy, et al., Assistance magnitude versus metabolic cost reductions for a tethered multiarticular soft exosuit, Sci. Robot. 2 (2) (2017) eaah4416.
[38] Z. Zake, F. Chaumette, N. Pedemonte, S. Caro, Vision-based control and stability analysis of a cable-driven parallel robot, IEEE Robot. Autom. Lett. (2019).
[39] D.A. Winter, Biomechanics and Motor Control of Human Movement, 9, John Wiley & Sons, Inc, Hoboken, NJ, 2009.
[40] R.A. Magill, Motor Learning and Control: Concepts and Applications, McGraw-Hill, 2011.
[41] S. Srivastava, P.-C. Kao, S.H. Kim, P. Stegall, D. Zanotto, J.S. Higginson, et al., Assist-as-needed robot-aided gait training improves walking function in individuals following stroke, IEEE Trans. Neural Syst. Rehabil. Eng. 23 (6) (2015) 956—963.
[42] J. Son, J. Ryu, S. Ahn, E.J. Kim, J.A. Lee, Y. Kim, Effects of 4-week intensive active-resistive training with an EMG-based exoskeleton robot on muscle strength in older people: a pilot study, BioMed. Res. Int. 2016 (2016) 1—5.
[43] L.A. Boyd, E.D. Vidoni, B.D. Wessel, Motor learning after stroke: is skill acquisition a prerequisite for contralesional neuroplastic change? Neurosci. Lett. 482 (1) (2010) 21—25.
[44] N. Lodha, P. Patel, A. Casamento-Moran, E. Hays, S.N. Poisson, E.A. Christou, Strength or motor control: what matters in high-functioning stroke? Front. Neurol. 9 (2018) 1160.

[45] L. Zhang, L. Li, Y. Zou, K. Wang, X. Jiang, H. Ju, Force control strategy and bench press experimental research of a cable driven astronaut rehabilitative training robot, IEEE Access. (2017).
[46] A.J. Young, H. Gannon, D.P. Ferris, A biomechanical comparison of proportional electromyography control to biological torque control using a powered hip exoskeleton, Front. Bioeng. Biotechnol. 5 (37) (2017).
[47] D. Ao, R. Song, J. Gao, Movement performance of human-robot cooperation control based on EMG-driven hill-type and proportional models for an ankle power-assist exoskeleton robot, IEEE Trans. Neural Syst. Rehabil. Eng. 25 (8) (2017) 1125−1134.
[48] F. Sylos-Labini, V.L. Scaleia, A. d'Avella, I. Pisotta, F. Tamburella, G. Scivoletto, et al., EMG patterns during assisted walking in the exoskeleton, Front. Hum. Neurosci. 8 (2014) 1−12.
[49] A.M. Fenuta, A.L. Hicks, Metabolic demand and muscle activation during different forms of bodyweight supported locomotion in men with incomplete SCI. BioMed Res. Int. 2014 (2014).
[50] Y. Ding, M. Kim, S. Kuindersma, C.J. Walsh, Human-in-the-loop optimization of hip assistance with a soft exosuit during walking, Sci. Robot. 3 (15) (2018) eaar5438.
[51] C. Walsh, Human-in-the-loop development of soft wearable robots, Nat. Rev. Mater. 3 (1) (2018) 5.
[52] X. Jin, A. Prado, S.K. Agrawal, Retraining of human gait - are lightweight cable-driven leg exoskeleton designs effective? IEEE Trans. Neural Syst. Rehabil. Eng. 26 (4) (2018) 847−855.
[53] N.C. Soderstrom, R.A. Bjork, Learning versus performance: an integrative review, Perspect. Psychol. Sci. 10 (2) (2015) 176−199.
[54] G. Kwakkel, Impact of intensity of practice after stroke: issues for consideration, Disabil. Rehabil. 28 (13−14) (2006) 823−830.
[55] S. Liu, D. Meng, L. Cheng, M. Chen, An iterative learning controller for a cable-driven hand rehabilitation robot, in: Proceedings IECON 2017 - 43rd Annual Conference of the IEEE Industrial Electronics Society, Institute of Electrical and Electronics Engineers Inc., January 2017, pp. 5701−5706.
[56] J.F. Veneman, R. Ekkelenkamp, R. Kruidhof, F.C.T. van der Helm, H. van der Kooij, A series elastic- and bowden-cable-based actuation system for use as torque actuator in exoskeleton-type robots, Int. J. Robot. Res. 25 (3) (2006) 261−281.
[57] H. Cao, D. Zhang, Soft robotic glove with integrated sEMG sensing for disabled people with hand paralysis, in: 2016 IEEE International Conference on Robotics and Biomimetics, ROBIO 2016, 2016.
[58] V. Vashista, X. Jin, S.K. Agrawal, Active Tethered Pelvic Assist Device (A-TPAD) to study force adaptation in human walking, in: Proceedings - IEEE International Conference on Robotics and Automation, 2014, pp. 718−723.
[59] R. Hidayah, S. Chamarthy, A. Shah, M. Fitzgerald-Maguire, S.K. Agrawal, Walking with augmented reality: a preliminary assessment of visual feedback with a cable-driven active leg exoskeleton (C-ALEX), IEEE Robot. Autom. Lett. 4 (4) (2019) 3948−3954.
[60] X. Jin, X. Cui, S.K. Agrawal, Design of a cable-driven active leg exoskeleton (C-ALEX) and gait training experiments with human subjects, in: Proceedings - IEEE International Conference on Robotics and Automation, June 2015, pp. 5578−5583.
[61] J. Kang, D. Martelli, V. Vashista, I. Martinez-Hernandez, H. Kim, S.K. Agrawal, Robot-driven downward pelvic pull to improve crouch gait in children with cerebral palsy, Sci. Robot. 2 (8) (2017) eaan2634.
[62] R. Hidayah, X. Jin, S. Chamarthy, M.M. Fitzgerald, S.K. Agrawal, Comparing the performance of a cable-driven active leg exoskeleton (C-ALEX) over-ground and on

a treadmill, in: Proceedings of the IEEE RAS and EMBS International Conference on Biomedical Robotics and Biomechatronics, August 2018, pp. 299–304.
[63] M. Khan, T. Luna, V. Santamaria, I. Omofuma, D. Martelli, E. Rejc, et al., Stand trainer with applied forces at the pelvis and trunk: response to perturbations and assist-as-needed support, IEEE Trans. Neural Syst. Rehabil. Eng. 27 (9) (2019) 1855–1864.
[64] M.I. Khan, V. Santamaria, S.K. Agrawal, Improving trunk-pelvis stability using active force control at the trunk and passive resistance at the pelvis, IEEE Robot. Autom. Lett. 3 (3) (2018) 2569–2576.
[65] G. Wulf, R. Lewthwaite, Optimizing performance through intrinsic motivation and attention for learning: the OPTIMAL theory of motor learning, Psychon. Bull. Rev. 23 (5) (2016) 1382–1414.
[66] M.I. Khan, V. Santamaria, J. Kang, B.M. Bradley, J.P. Dutkowsky, A.M. Gordon, et al., Enhancing seated stability using trunk support trainer (TruST), IEEE Robot. Autom. Lett. (2017).
[67] M.I. Khan, A. Prado, S.K. Agrawal, Effects of virtual reality training with trunk support trainer (TruST) on postural kinematics, IEEE Robot. Autom. Lett. 2 (4) (2017) 2240–2247.
[68] V. Vashista, D. Martelli, S.K. Agrawal, Locomotor adaptation to an asymmetric force on the human pelvis directed along the right leg, IEEE Trans. Neural Syst. Rehabil. Eng. (2016).
[69] J. Kang, K. Ghonasgi, C.J. Walsh, S.K. Agrawal, Simulating hemiparetic gait in healthy subjects using TPAD with a closed-loop controller, IEEE Trans. Neural Syst. Rehabil. Eng. (2019).
[70] D. Martelli, J. Kang, S.K. Agrawal, A perturbation-based gait training with multidirectional waist-pulls generalizes to split-belt treadmill slips, in: Proceedings of the IEEE RAS and EMBS International Conference on Biomedical Robotics and Biomechatronics, 2018.
[71] L. Bishop, J. Stein, V. Vashista, M. Khan, S. Hinds, S. Agrawal, Stroke survivor gait adaptations using asymmetric forces with the tethered pelvic assist device, Arch. Phys. Med. Rehabil. 96 (12) (2015) e20.
[72] L. Bishop, The Integration of Principles of Motor Learning to Reduce Gait Asymmetry Using a Novel Robotic Device in Individuals Chronically Post-Stroke (Ph.D. thesis), Columbia University, 2018.
[73] D. Martelli, F. Aprigliano, S.K. Agrawal, Gait adjustments against multidirectional waist-pulls in cerebellar ataxia and parkinson's disease, International Conference on NeuroRehabilitation, Springer, Cham, 2019, pp. 283–286.
[74] V. Vashista, N. Agrawal, S. Shaharudin, D.S. Reisman, S.K. Agrawal, Force adaptation in human walking with symmetrically applied downward forces on the pelvis, in: Proceedings of the Annual International Conference of the IEEE Engineering in Medicine and Biology Society, EMBS, 2012.
[75] F. Alton, L. Baldey, S. Caplan, M.C.C. Morrissey, A kinematic comparison of overground and treadmill walking, Clin. Biomech. (Bristol, Avon) 13 (6) (1998) 434–440.
[76] R. Hidayah, Gait adaptation using a cable-driven active leg exoskeleton (C-ALEX) with post-stroke participants, IEEE Trans. Neurosci. Eng. 28 (9) (2020) 32746320.
[77] M. Casadio, V. Sanguineti, Learning, retention, and slacking: a model of the dynamics of recovery in robot therapy, IEEE Trans. Neural Syst. Rehabil. Eng. 20 (3) (2012) 286–296.

CHAPTER 6

XoSoft: design of a novel soft modular exoskeleton

Jesús Ortiz, Christian Di Natali and Darwin G. Caldwell
Department of Advanced Robotics, Italian Institute of Technology, Genova, Italy

6.1 Introduction

A relevant amount of the population, such as elderly and patient groups, suffer from mobility impairments of varying degrees. The world's elderly population (over 60 years old) is expected to increase significantly by 2050, reaching about 2 billion and doubling the proportion from 2000 [1]. Aging is associated with a decline in voluntary muscle strength and important changes to body composition and function [2]. In order to slow down the degrading of overall health and cognitive functions, it is critical to remain active and mobile [3]. Patient groups, such as poststroke or incomplete spinal cord injury (SCI) patients also require mobility assistance. Around 16 million people per year suffer a stroke for the first time, and an important percentage of them (c. 5 million) experience varying degrees of mobility difficulty, which significantly impacts their ability to perform tasks of daily living [4]. SCI are mostly caused by traumatic accidents, of which around half are incomplete (the patient suffers from a partial function loss) [5]. Apart from these cases, there is an increasing number of other different causes, mostly affecting elderly people, in this case with a higher percentage of incomplete SCI. Incomplete SCI does not produce a complete loss of sensory–motor function in the lower limbs, but the partial loss of mobility still may require assistance to walk.

Most of these population groups make use of assistive devices, which allow them to keep some degree of independency and perform basic tasks of daily living. However, many of these devices simply substitute or complement some function loss, but do not encourage the activation or rehabilitation of legs. For example, powered wheelchairs do provide mobility, but they do not help to exercise the legs, and consequently contribute to the functionality loss.

Active exoskeleton devices have the ability to assist the user, but at the same time keep their mobility contributing to the preservation and regeneration of their residual functionalities [5]. Some devices target only the clinical rehabilitation phase, and are based on fixed platforms, such as Lokomat [6,7] or LOPES [8]. While these devices offer a very good tool for rehabilitation, they cannot be used in daily life activities. Other kinds of lower limb exoskeletons replace a complete loss of mobility, for example ReWalk [9], Ekso [10], or Indego [11]. While these devices are an alternative to a wheelchair for paralyzed people, they are too complex and expensive for patients with a low to moderate degree of impairment.

A new generation of exoskeletons based on soft technologies are in development with great promise in terms of usability and performance. They target specifically patients that still keep some degree of mobility, and consequently the exoskeleton only provides a partial support, which is suitable for the current soft technologies, for example, the Harvard soft exosuit [12], or MyoSwiss [13]. Both systems share common elements, such as cable-driven actuation, but with different approaches in the implementation.

XoSoft EU Project targeted the development of a new generation of lower-limb modular soft exoskeleton for the assistance of people with mobility restrictions due to a partial loss of sensory or motor function [14,15]. One of the main challenges of the project was to develop a complete soft system, including smart soft sensors and actuators. This chapter includes a full picture of the development of the XoSoft project. The user-centered design (UCD) approach, described in Section 6.2. Section 6.3 presents the preliminary data collection from the different user groups, which was used for the definition of the system requirements. Sections 6.4 and 6.5 present the actuation and sensing technologies, while Section 6.6 describes each prototype developed during the project. The testing and validation of the system is described in Section 6.7. The conclusions of this chapter are presented in Section 6.8, including also future works.

6.2 User-centered design

XoSoft has been developed according to the UCD approach [16,17]. UCD, or human-centered design, is defined in ISO 9241-210 as "an approach to interactive systems development that aims to make systems usable and useful by focusing on the users, their needs and requirements, and by applying human factors/ergonomics, and usability knowledge and techniques" [18]. UCD directly involves the stakeholders in all the stages of development to ensure a clear understanding of the requirements and guarantee that the design is driven and refined by user-centered evaluation. This is achieved by involving the users from the earliest stages of the development (using alpha prototypes) and following an iterative design, evaluation, and redesign process. The main purpose of this approach to the design of an exoskeleton is (1) to ensure the effectiveness and efficiency of the system; (2) to improve user acceptance; (3) to reduce safety risks and other adverse effects; (4) to ensure continued use of the system; and (5) to enhance user well-being.

The main XoSoft stakeholders can be divided in three different user groups:

1. *Primary users (PU)*. PUs are the users that wear the system and get the most direct benefit from it. XoSoft targets patients with low to medium mobility impairments, in particular two distinct groups: (1) neurological populations, specifically people with stroke and incomplete SCI, and (2) older adults.
2. *Secondary users (SU)*. SUs are any person or organization directly in contact with a PU. We can divide them in two groups: (1) professional, and (2) nonprofessional. Professional SUs include healthcare professionals (e.g., physicians, nurses, physiotherapists, occupational therapists, speech and language therapists, and public health nurses), as well as professional care assistants, home help service providers, and other support staff. Nonprofessional SUs may include spouses, family members, friends, neighbors, and community and/or voluntary organizations.
3. *Tertiary users (TU)*. Any other user that has any interest that affects the potential for the device in question is a TU. They are often related to the framework that is in place on the regulatory side and/or to financial provisions for the devices that are supplied to patients or other users.

Mainly, PUs and SUs have been involved in the initial analysis of the user requirements of the target system. The main findings of this study are described in Section 6.3.

The iterative approach adopted in this work emphasizes the development of several prototypes that will be tested and evaluated by different user groups. XoSoft includes the development of four distinct prototypes, as shown in Fig. 6.1:

1. *Alpha version*. This version makes use of existing technologies, and instead of a single prototype was formed from separate modules

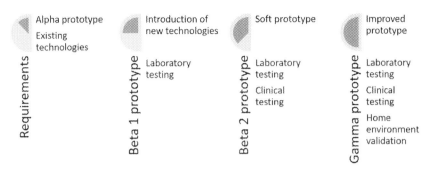

Figure 6.1 UCD approach followed in the development of XoSoft.

representing different functionalities of the system: (1) capacitive knee soft sensor, (2) capacitive ankle soft sensor, (3) three different resistive knee soft sensors, (4) variable stiffness knee hard exoskeleton, (5) variable stiffness ankle hard exoskeleteon, (6) sensorized sock for estimation of ground contact, and (7) two garment concepts. Its main purpose is to evaluate different possible technologies and concepts to integrate into the subsequent prototypes, and to get initial feedback from the different user groups and start the involvement of the users in the design process.

2. *Beta 1 and 2 versions*. These are the first prototypes incorporating new technologies into the system. During the development of XoSoft, two iterations of the Beta prototype were developed. This allowed testing of actuation and sensing principles, without incorporating too many new changes at the same time. The Beta 1 was the first soft prototype, but still made use of existing technologies for actuation and sensing. This prototype was useful to understand the limitations and potential of a soft exoskeleton. The Beta 2 prototype already incorporated most of the novel technologies for sensing and actuation, but still there were margins for improvement in terms of autonomy, weight, and performance.

3. *Gamma versi*. This is the final prototype of the system, which is heavily based on the Beta 2 prototype but improves all the technical aspects to increase performance and robustness. This is essential for the final validation of the system, where the reliability of the prototype has an important impact on the user acceptance evaluation. The main improvements were on the garment, low-level control for better energy efficiency, improved electronics, improved actuators, and the onboard energy system.

The technical details of the Beta 1, Beta 2, and Gamma prototypes are fully described in Section 6.7.

The testing and validation of the different prototypes is divided into three different phases, which are detailed in Section 6.8:

1. *Laboratory testing*. The purpose of this testing is to perform a biomechanical study using standard gait analysis equipment (force plate, motion tracking, and electromyography—EMG). This gives quantitative information about the evolution of the different prototypes. The Beta 1, Beta 2, and Gamma prototypes were all tested in this setup.
2. *Clinical testing*. In this phase, the prototypes were tested in clinical conditions, similar to those that a patient will encounter during rehabilitation. In this case, the information acquired is qualitative using metrics similar to those used in a standard clinical evaluation of a patient (gait parameters, walking index, etc.). Only the Beta 2 and Gamma prototypes were tested at this stage. Being in a clinical environment, it is important that the prototype is robust enough. The testing with an early prototype would not give very meaningful information and it could discourage the patients.
3. *Validation in home-simulated environments*. Finally, in order to understand how the users could use the system in their daily life, the Gamma prototype was tested in a simulated home environment. The choice of a simulated environment instead of a real scenario was made to ensure that it was possible to have continuous monitoring of the situation to collect representative data and avoid safety issues.

6.3 Requirements

At the beginning of the project, a cross-sectional mixed-methods study was carried out involving PUs and SUs (as previously defined) in order to define prioritized user requirements [19]. This was very valuable in helping to understand the most important features that should be included in XoSoft, and to separate these critical features from the desirable characteristics. Obviously, each user group has different priorities, but the combination of PUs and SUs requirements offered a quite complete list of system specifications. In a first campaign, six stroke/SCI patients, six frail adults, and three pre-frail adults were included in the PUs group. They were from three different European cities in Germany, Ireland, and The Netherlands to avoid bias due to the country of origin. The SU group included 26 participants from four different European cities in Germany,

Ireland, The Netherlands, and Switzerland, and consisted of healthcare professionals, and nonprofessional caregivers (family and friends of PUs). The interviews were designed specifically for each user group, since each of them might have different priorities.

6.3.1 Primary users

Most of the PUs were very positive about the XoSoft concept and the potential benefit in terms of mobility assistance. The hands-free usage of the system was seen as a distinct benefit with respect to other assistive devices. Poststroke patients felt that XoSoft could help them with the foot drop problem (inability or impaired ability to raise the toes or raise the foot from the ankle—dorsiflexion). Frail adults were less positive about the use of XoSoft and did not perceive a great potential benefit of the system because they felt it was too complex and difficult to use, becoming an unnecessary burden to them. The main requirements gathered from the PUs were:

- *Benefits and functions.* The highest priority was given to the effectiveness of the system as an assistive device. In particular, a device that could provide (1) better quality walking for longer distances with less effort; (2) adaptable support for lower body joints, as required; (3) help and support with bending activities; and (4) hands-free solution. Safety was also an important aspect from two different points of view: safe ambulation and safe use in the presence of pathologies such as cardiovascular diseases or skin conditions.
- *Practicability.* The system should be easy to use and wear (considering that some of the patients might have movement limitations also in the upper body, e.g., poststroke patients) and easy to store. It should be also lightweight, but robust enough to provide the required support.
- *Training and support.* Considering the age of the potential users, their main concern was related to the learning phase. For the case of poststroke and SCI patients, they would not like to spend too much time learning how to use the system, but they would accept some assistance from another person.
- *Design and* ae*sthetics.* The opinion on the aesthetics was mixed. Some of the PUs considered the aesthetics of the device as a secondary priority with respect to the functionality, while others gave particular importance to this aspect. One common opinion was that the system

should not be too bulky and that it should be possible to wear it below their usual clothes.
- *Costs.* Either an affordable device at a low price, or with a reimbursement from health insurance companies was the preference of the PUs.

6.3.2 Secondary users

The general opinion of the SUs was that the system could be potentially used by many different patients thanks to the flexibility and adaptive nature of the XoSoft concept. However, in the case of neurological rehabilitation applications, they indicated that there are already a number of available exoskeleton products. Nevertheless, a device that could be used outside a clinical setup, with the appropriate characteristics in terms of usability, would be very beneficial and it would offer more than any existing device. The main requirements pointed out by the SUs were:

- *Benefits and functions.* In this case, the priority was put into safety, including fall prevention. However, the mobility assistance and the improvement of the independence of the PUs was also an important aspect.
- *Practicability.* The SUs also indicated the easiness to use as an important aspect, highlighting that it should be usable by patients with cognitive and upper body impairments.
- *Training and support.* SUs indicated that the system should be usable without a large amount of training and support, otherwise this would limit the usage and acceptance by both the PUs and SUs. They also pointed out that this should be part of the overall product service.
- *Design and aesthetics.* It was considered that the aesthetics could play a different role depending on the scenario. Aesthetics when used in rehabilitation applications was not considered a priority, however, when applied in community/social settings appearance was seen to be much more important, mostly to avoid potential stigmas. Other characteristics that were indicated from the design point of view were light weight, robustness, ease of cleaning, compatibility with usual footwear and clothing, and not being bulky.
- *Costs.* The opinion of the SUs was compatible with that of the PUs, highlighting that for expensive devices reimbursement would be critical to facilitate uptake.

6.4 Actuation principle

One of the main characteristics of XoSoft is the use of a quasipassive actuation principle. This means that the system does not assist the user through an actuator of any kind (electric motor, pneumatic/hydraulic system, etc.), but uses passive elements that can store and release mechanical energy, regulated by an active component. This actuation principle was chosen from a preliminary assessment [20] in which, through an analysis of the gait of a sample patient, the combination of specific passive and active elements showed important potential benefits in terms of peak torque and power, as well as energy involved during the whole gait cycle. This has several advantages: (1) there is no need for a high-power actuation system; (2) the energy efficiency of the system (and consequently the autonomy) can be substantially improved; and (3) this gives opportunities to explore alternative actuation mechanisms that are more suitable for a soft system. It is worth noting that while the healthy human gait is highly optimized with only a small margin for improvement, for patients with a small/medium degree of disability, a quasipassive actuation system has the potential for significant improvement.

During the development of XoSoft, different implementations of quasipassive actuation systems were considered, as described in the following subsections.

6.4.1 Textile jamming

The granular jamming principle is a well-known mechanism used in soft robotics [21] to construct soft variable stiffness mechanisms. A closed flexible (elastic) bag filled with beads of different nature (coffee powder, polymeric beads, etc.) can modify considerably its stiffness by changing the internal pressure. Under normal conditions (internal pressure equal to the external one), the bag can be easily deformed presenting a low stiffness. However, by applying a vacuum the beads collapse presenting high internal forces, and consequently increasing the stiffness of the whole structure. During XoSoft, a modification of this principle was used to develop a textile jamming mechanism. In the XoSoft system, the beads are replaced by a textile where the fibers are doing the same function. The advantage of this system is that the textile is a structured material and ensures a better distribution of the fibers along all the volume. The practical implementation of this system was functional

Figure 6.2 Textile jamming implementation on a rectangular mat. The prototype was able to vary the stiffness with the change of internal pressure. Full video at https://www.youtube.com/watch?v = jMj-TlbB08A.

with the required stiffness variation (see Fig. 6.2), however due to the high internal volume and actuation time (over 5 s), it was not practical for the target application.

6.4.2 Electromagnetic clutch

In the study carried out by Ortiz et al. [20], a mechanism that has the potential to provide a good level of assistance can be produced by an elastic element in series with a clutch, that is, a mechanism that is able to behave both as a transparent (without resistance to the motion) and/or a stiff element. With the first (transparent) stage, the user perceives only minimally the presence of the actuation element, while in the second (stiff) stage it is possible to store and release mechanical energy, providing assistance to the user. Practical implementation of this mechanism is, however, very challenging in a soft exoskeleton. The first implementation in XoSoft (Beta 1 prototype) made use of existing technologies in combination with a Bowden cable transmission to transfer the assistance to the target joint while retaining the soft nature of the system. The elastic component was a rubber band mounted directly into the joint, making use of the soft nature of this material, as opposed to a more rigid metal spring. The clutch was implemented using a commercial electromagnetic clutch with a custom assembly to transform from rotary to linear motion. Fig. 6.3 shows the implementation of the clutch. Note that there is also a retracting mechanism using an elastic band and a pulley, that helps to return the system to the neutral position. While, this is necessary in order to avoid any slack of the mechanism it also introduces a small resistance in the system.

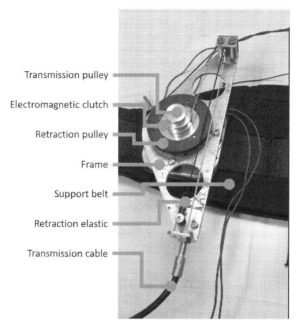

Figure 6.3 Implementation of the electromagnetic clutch with the connection to the Bowden cable and the retraction mechanism.

6.4.3 Pneumatic soft clutch

Combining the principle of the granular and textile jamming with the functionality of the electromagnetic clutch, it is possible to construct a soft linear clutch to be mounted in series with the elastic element. From the functional point of view, this is equivalent to the electromagnetic clutch mechanism, but it can be mounted directly on the human joint due to the flexible nature of the pneumatic systems, with important advantages from the wearability point of view. The pneumatic soft clutch is fully described in [22], and it can be seen in Fig. 6.4. It is composed of (1) two textile straps with embedded 3D plastic structures (in order to increase the friction forces); (2) two retracting elastic bands; and (3) a latex-made elastic cover to enclose the whole structure and ensure the proper generation of the vacuum. In the natural condition (internal pressure is equal to the external one), the two textile straps can slide freely. When a vacuum is applied, the atmospheric pressure exerts a pressure on the textile straps increasing the friction between them, and consequently blocking the motion up to 100 N of traction force. This, effectively, creates a soft

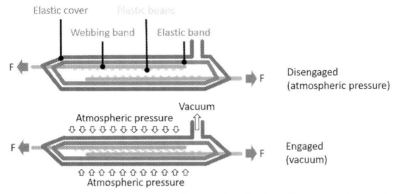

Figure 6.4 Pneumatic soft clutch. On the top, the clutch is disengaged (atmospheric pressure) and the resistance force is low. On the bottom, the clutch is engaged (vacuum) and the resistance force is high.

component that is able to change from a free motion (with the exception of the force created by the retracting elastic straps) to a blocking status.

6.5 Sensing and control

Considering the nature of the quasipassive actuation, the main information required by the control system is the gait phase, since the clutches only have two states and they are only engaged and disengaged once per cycle. Consequently, the core of the control system is a state machine that keeps track of the gait of the patient. The gait can be divided into two primary phases: (1) stance, when the corresponding foot is touching the ground; and (2) swing, when the foot is in the air. The transition between both phases is marked by the heel-strike (HS) and by the toe-off (TO), respectively. The gait cycle is considered to start at the HS (0%) and the TO indicates the 60% point of the gait phase. This is a standard distribution of the gait that normalizes in time the gait movement. In order to refine this segmentation, it is common to consider additional events. The full list of considered events is listed below:

1. *Heel-Strike (HS)*. When the heel touches the ground. It marks the start of the stance phase of the leg.
2. *Foot-Flat (FF)*. The whole foot is touching the ground.
3. *Mid-Stance (MS)*. When the opposite leg crosses the support leg.
4. *Heel-Off (HO)*. When the heel loses contact with the ground.

5. *Toe-Off (TO)*. When the toe of the foot loses contact with the ground. This marks the start of the swing phase, since the foot is completely in the air.
6. *Mid-Swing (MSW)*. When the leg crosses the opposite (support) leg. It is the equivalent of the midstance of the opposite leg.

To implement this action some sensor information is required to coordinate the motion. In particular, the XoSoft prototype makes use of different sensory information:

- Contact between the feet and the floor. This information is useful to identify the HS, FF, HO, and TO events and to be able to segment the stance phase.
- Angle/velocity of the knee. To segment the swing phase, it is possible to use the information relative to the angle or velocity of the knee.

Once the gait has been segmented, it is possible to estimate the percentage of the gait cycle and engage/disengage the clutches according to some predefined criteria, that can be (1) percentage of the gait cycle, or (2) a delay with respect to a specific event (from the list above). The selection of the criteria and the percentage/delay is made according to an evaluation of the patient's gait by an expert clinician.

In the following subsections, the different sensor approaches used in XoSoft are described.

6.5.1 Insole pressure sensors

As described in the previous subsection, it is important to measure the contact between the foot and the floor to identify different events during the stance phase. Considering this, XoSoft includes a sensorized insole with pressure sensors placed in strategic places. The distribution of the sensors, as shown in Fig. 6.5, is (1) rear sensor placed approximately under the heel, (2) middle sensors placed between the heel and the toes, and (3) front sensor placed at the tip of the toes. In XoSoft, these sensors are commercial force sensing resistors (FSR), which measure the contact force transmitted through their surface. These sensors are integrated into a custom-made silicon insole that can be used in any shoe. Since the measurement provided by the FSR sensors is not very reliable, not only because of the sensor itself, but also because of the force distribution along all the foot surface, it is not possible to use the magnitude of the force. However, in order to capture the gait events, only a binary on/off signal is required, and this can be extracted with a simple threshold operation.

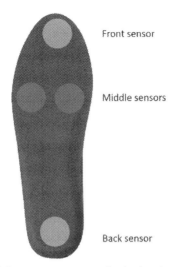

Figure 6.5 Placement of the pressure sensors in the insole.

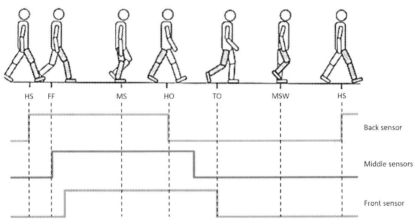

Figure 6.6 Gait phases and activation of the insole pressure sensors of the right foot.

The signal provided by the insole pressure sensors can be easily understood in Fig. 6.6, showing the triggering signal together with the gait phases. The HS event is directly given by the rising edge of the rear sensor signal, the FF occurs on the rising edge of the middle sensor signal, the HO is at the falling edge of the rear sensor signal, and TO is indicated by the falling edge of the front sensor.

6.5.2 IMUs

In order to estimate the knee angle and velocity, it is possible to use a combination of two inertial measurement unit (IMUs) placed on the shank and thigh, respectively. As we can see in Fig. 6.7, the knee angle changes direction at around 70% of the gait cycle, with a short advance with respect to the MS1 event. This event can be measured by estimating the knee angle or velocity. With a proper calibration it is possible to calculate the angle difference between the two IMUs and estimate the knee angle. However, since the IMUs make use of a magnetometer to calculate the absolute rotation matrix, they are very sensitive to electromagnetic noise, offering a low reliability in unstructured scenarios. For this reason, the knee velocity was used instead as it does not need information from the magnetometer, however, on the downside it does require a more elaborate calibration [23].

6.5.3 Soft sensors

XoSoft includes also an alternative way of estimating the knee angle using novel soft sensors integrated into the knee garment. This offers a more integrated solution with respect to the use of IMUs, but due to the maturity of this technology, they do not offer the same level of robustness

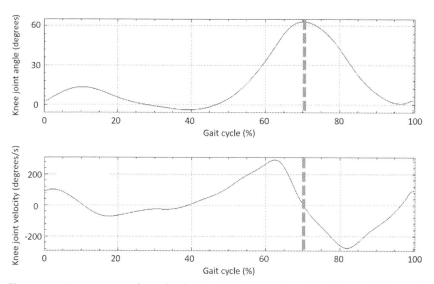

Figure 6.7 Knee joint angle and velocity during normal walking. The angle is maximum and the velocity crosses zero at around 70% of the gait.

from the physical reliability point of view. However, they offer some important advantages, such as better integration, immunity against electromagnetic disturbances, and easier calibration. The disadvantage is that the wrinkling of the garment can significantly affect the reading of the sensors, making it impossible to differentiate small knee angles. However, since they are mostly used for the swing phase where the knee angle motion is relatively large, they still provide very useful information.

From the technical point of view, the sensors integrated in XoSoft are capacitive sensors made of several layers of dielectric and nondielectric material [24]. The main principle is depicted in Fig. 6.8. In practice, bending of the knee decreases the distance between the layers, this leads to a variation of the capacitance. As described before, for low flexion angles (or hyperextension) after a big flexion, the textile can present some wrinkling.

6.6 Prototypes
6.6.1 XoSoft Beta 1

The Beta 1 prototype (Fig. 6.9) was the first design to include sensing and actuation in a soft structure. At that stage of the development cycle, the required technologies for soft actuation were not at the required maturity level to integrate them into the prototype. For this reason, we developed an equivalent actuation system, based on the same principle of quasipassive actuation, but using existing technologies, that is, with an electromagnetic clutch as described in Section 6.4.2. To simplify the design of the Beta 1 prototype, we focused on a single leg, to be used to assist unilateral patients, and only two joints (hip and knee). In particular, the Beta 1 prototype was designed to assist the hip and knee flexion, with the aim of

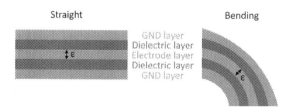

Figure 6.8 Capacitive soft sensor. On the left, the sensor is straight, and the electrode layer has a thickness ε. On the right, the sensor is bent, and the thickness ε of the electrode is reduced, producing a variation of the capacitance between the two dielectric layers.

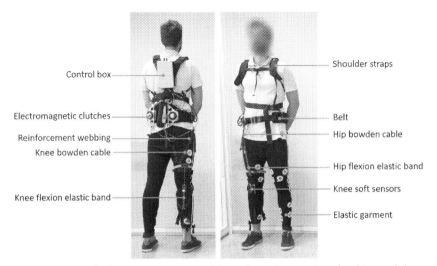

Figure 6.9 XoSoft Beta 1 prototype with unilateral actuation for hip and knee flexion.

improving the foot clearance and other functional walking parameters (symmetry, etc.).

From the sensory information point of view, this prototype incorporated the insole pressure sensors and the soft sensors integrated into the knee. However, only the foot sensors were integrated into the control, with the knee sensors only being used to estimate the knee angle. This prototype also incorporated a first implementation of the electronics architecture, including all the secondary boards for the real-time communication with the foot sensors and knee sensors. More details of this prototype can be found in [25].

6.6.2 XoSoft Beta 2

The XoSoft Beta 2 prototype (Fig. 6.10) represented an important leap forward since it included the quasipassive actuation based on the pneumatic soft clutches, as described in Section 6.4.3. This improved significantly the wearability of the system, complying with the objective of a full soft exoskeleton. It included an improved version of the soft capacitive sensors integrated in a knee garment, and IMUs for the segmentation of the swing gait phase. The garment used in the system offered a number of important improvements that allowed the actuation of hip, knee, and ankle in both directions (flexion/extension and plantar/dorsi flexion,

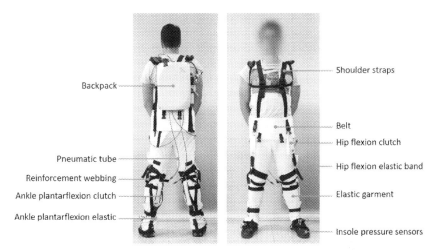

Figure 6.10 XoSoft Beta 2 prototype configured with a bilateral actuation for hip flexion and ankle plantarflexion.

respectively) on both legs. The total number of actuation locations were 12, but due to limitations in the pneumatic circuit, the number of simultaneous actuators was limited to eight. However, none of the configurations used during testing was higher than six, since some combinations were found to be either unfeasible or not useful. The electronics of the system was completely updated, since it needed to include all the elements required for the pneumatic control, that is, pressure sensors, valve control, etc. All this equipment was placed on a bigger backpack, with a weight of around 4 kg. This prototype required an external electrical power supply and an external air pressure source.

6.6.3 XoSoft Gamma

The XoSoft Gamma prototype (Fig. 6.11) was the final prototype with only a few but important changes with respect to the Beta 2 prototype. The main difference was the garment itself, which was a completely new design based on an upgraded approach. This redesign was the result of the UCD approach building on the testing and experience gained with the Beta 1 and Beta 2 prototypes. In practice, the Beta prototypes used a tight garment, which ensures a good fit to the subject. However, each subject needs their own personalized garment. While this was feasible in the early stages of the project where the prototypes were tested with a few subjects, it was not a practical solution for the testing and validation of the Gamma

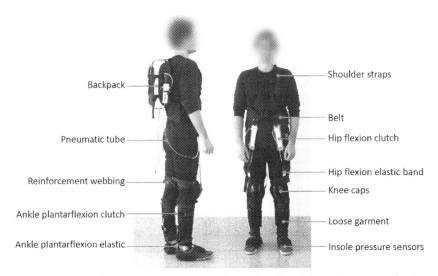

Figure 6.11 XoSoft Gamma prototype configured with a bilateral actuation for hip flexion and ankle plantarflexion.

prototype with a larger pool of users. Also, an eventual future commercial implementation of the system was taken into consideration. For this reason, the garment of the XoSoft Gamma prototype was based on loose trousers with webbing elements for the fixation to the body. Furthermore, the use of a loose garment also allowed a quicker donning and doffing.

The number of simultaneous actuators was reduced to six, with the corresponding saving of weight and space in the backpack, which allowed the placement of a battery. The total weight of the backpack remained similar (approx. 4 kg). The Gamma prototype was prepared to work with an optional pressure tank, in order to make the system completely untethered.

The low-level control of the valves was improved in order to reduce the air flow consumption. This reduced the energy consumption to around one third, increasing significantly the autonomy of the system.

This device was certified as a medical device Class I, in conformity with the standard UNI ISO 60601. More details of this prototype can be found in [26].

6.6.4 Comparison

To better understand the differences between the various prototypes, Table 6.1 shows a schematic of the main characteristics of each device.

Table 6.1 Comparison between XoSoft prototypes.

	XoSoft Beta 1	XoSoft Beta 2	XoSoft Gamma
Target	Unilateral hip and knee	Bilateral hip, knee, and ankle	Bilateral hip, knee, and ankle
Garment	Tight garment with hip and knee attachments	Tight garment with hip, knee and ankle attachments	Loose fitting garment with hip, knee and ankle attachments, backpack integrated support
Actuation principle	Quasipassive, variable stiffness	Quasipassive	Quasipassive
Actuation technology	Electromagnetic clutches	Soft pneumatic clutch	Soft pneumatic clutch
Number of actuators	2	8	6
Insole pressure sensors	4 FSR sensors per foot	4 FSR sensors per foot	4 FSR sensors per foot
IMUs	2 per leg (not for control)	2 per leg	2 per leg
Soft knee sensors	Yes (not for control)	Yes (not for control)	Yes
Low-level control	On/off control	On/off pneumatic control	Optimized control
High-level control	FSM based on foot sensors	FSM based on foot sensors and IMUs	FSM based on foot sensors and IMUs or knee sensors
Energy supply	Onboard battery	External electrical. External pneumatic	Onboard battery. External pneumatic. Onboard pressure tank
Monitoring and feedback	No	Preliminary	Integrated
Certification	No	No	Medical Device Class I

6.7 Testing and validation

As already introduced in Section 6.2, each prototype was tested in different conditions. Table 6.2 shows the different testing phases carried out with each prototype. The laboratory testing was carried out with all the prototypes. This was important to assess the improvements at each stage of the development cycle. The Beta 1 prototype was not suitable for testing with a large pool of subjects, since the robustness was insufficient, and this could create a rejection from the subjects. For this reason, only the laboratory testing was carried out. The Beta 2 prototype however, was a more robust prototype and it could be tested with seven elderly and frail adults. Finally, the Gamma prototype represented an important step in user testing and acceptance, and it was validated in a home-simulated environment.

The following sections describe the experimental protocols in more detail and highlight the main results arising from each of the testing phases.

6.7.1 Laboratory testing

All three XoSoft prototypes (Beta 1, Beta 2, and Gamma) were tested in laboratory conditions. The main purpose of this testing was to perform a comparative evaluation of each iteration of the prototype. The evaluation included general performance of the system, quantitative evaluation of the effects of XoSoft on the patient, as well as qualitative feedback from the users. The next section describes only the most relevant aspects of the experimental protocol used for the laboratory testing. More details can be found in [25,27,28].

Table 6.2 Testing and validation stages of each XoSoft prototype.

	XoSoft Beta 1	XoSoft Beta 2	XoSoft Gamma
Laboratory testing	1 unilateral PS	1 unilateral PS 3 iSCI	1 unilateral PS 3 iSCI
Clinical testing	—	7 EF	6 EF
Home-simulated environment	—	—	3 unilateral PS 2 iSCI

EF, Elderly/frail; *iSCI*, incomplete Spinal Cord Injury; *PS*, Poststroke.

6.7.1.1 Experimental protocol

The trials included different tasks performed under different conditions, including no exoskeleton and different configurations of the actuation and control. Three tasks were considered in the laboratory testing:
- *Straight walking.* The subjects were asked to walk in a straight line for approximately 10 m at a self-selected speed. The speed was maintained during all the trials and conditions. Each trial was repeated 10 times for each condition.
- *90 degrees turn.* The subjects performed a walking task with a 90-degree turn. Each trial was repeated 10 times for each condition.
- *Donning/doffing.* The subjects were asked to don and doff the system without the help of another person. This trial was only performed once per subject.

During the trials, different measurements were collected. A visual tracking system with 12 cameras (Vicon Vantage V5, Vicon Motion Systems Ltd, Oxford, United Kingdom) was used to record the motion of the subjects. Three or four retro-reflective markers were attached to each of the relevant body segments (forefoot, rearfoot, shank, thigh, pelvis, trunk). Ground reaction forces (GRF) were recorded with two floor-mounted AMTI force plates (AMTI Inc., Watertown, United States). Surface electromyography (sEMG) was recorded in the relevant muscle groups with FREEEMG (BTS Bioengineering Corp. Quincy, United States).

During the walking trial (straight or 90 degrees), the subjects were asked to walk on the plates, but in order to not disturb the walking pattern, they were not asked specifically to touch each plate with their left and right feet, respectively. Consequently, the joint torque analysis was only performed in a few trials where the GRF information was available. The motion data, however, was available for all 10 trials, except for missing information due to occlusions. This allowed a more detailed statistical analysis. To compare the results, all the data was normalized from 0% to 100% of gait cycle (0% = HS, 100% = consecutive HS).

The evaluation included also qualitative information recorded in the following way:
- *Rating of perceived exertion/pressure.* Subjects were asked to evaluate the perceived exertion and pressure of the device in 12−14 different body areas (see Table 6.3), using the CR10 Borg scale [29] (see Table 6.4).

Table 6.3 Body regions for the evaluation of the perceived exertion/pressure.

Index	Description	Index	Description
1	Right foot	8	Left foot
2	Right ankle	9	Left ankle
3	Right lower leg	10	Left lower leg
4	Right knee	11	Left knee
5	Right upper leg	12	Left upper leg
6	Right hip	13	Left hip
7	Right shoulder[a]	14	Left shoulder[a]

[a]Evaluated only in Beta 2 and Gamma prototypes.

Table 6.4 Borg scale for the categorization of the perceived exertion/pressure.

Rating	Description	Detailed description
0	Nothing at all	
0.5	Extremely weak	Just noticeable
1	Very weak	
2	Weak	Light
3	Moderate	
4	Somewhat strong	
5	Strong	Heavy
6		
7	Very strong	
8		
9		
10	Extremely strong	Almost maximal
–	Absolute maximum	Highest possible

Some of the values also include a description to help the subjects to select the perceived value.

- *System usability*. The prototypes were also evaluated using the standard system usability scale (SUS). This tool includes 10 questions, which were evaluated using a Likert scale from 1 (strongly disagree) to 5 (strongly agree) (see Table 6.5). There are five positive statements (1, 3, 5, 7, and 9) and five negative statements (2, 4, 6, 8, and 10). The positive statements value is the "score minus 1", and the negative statements value is "5 minus the score". The sum of all the scores is multiplied by 2.5, giving an overall value between 0 and 100, which is the

Table 6.5 System usability scale (SUS) questions and scoring system.

Statement	Strongly disagree	Score			Strongly agree	Value
1. I think that I would like to use this system frequently	1	2	3	4	5	$S_1 = score - 1$
2. I found the system unnecessarily complex.	1	2	3	4	5	$S_2 = 5 - score$
3. I thought the system was easy to use.	1	2	3	4	5	$S_3 = score - 1$
4. I think that I would need the support of a technical person to be able to use this system.	1	2	3	4	5	$S_4 = 5 - score$
5. I found the various functions in this system were well integrated.	1	2	3	4	5	$S_5 = score - 1$
6. I thought there was too much inconsistency in this system.	1	2	3	4	5	$S_6 = 5 - score$
7. I would imagine that most people would learn to use this system very quickly.	1	2	3	4	5	$S_7 = score - 1$
8. I found the system very cumbersome to use.	1	2	3	4	5	$S_8 = 5 - score$
9. I felt very confident using the system.	1	2	3	4	5	$S_9 = score - 1$
10. I needed to learn a lot of things before I could get going with this system.	1	2	3	4	5	$S_{10} = 5 - score$
System Usability (0–100) =						$\sum_{i=1}^{10} S_i \cdot 2.5$

System Usability. This metric is useful for comparing the different versions of the XoSoft prototype.

From the kinematic data it is possible to extract some metrics that can give quantitative information about the effects of the exoskeleton on the person [27]. The following values are calculated as the average of all the steps of a single trial:

- *Step length asymmetry.* This value gives an indication of the symmetry of the gait by comparing the step length of the left and right leg. This coefficient is more relevant for unilateral patients (typically poststroke patients). A value of 0 means that the step length is equal between the left and right leg, while a lower or higher value indicates some degree of asymmetry. Consequently, a positive effect of the exoskeleton should bring this value closer to 0. It is defined as:

$$SLA_k = 1 - \frac{x_u}{x_a}$$

where x_u is the step length of the unaffected leg, while x_a is the step length of the affected one.

- *Stride length.* The stride length is the distance covered by the left and right feet during a full gait. A pathological gait will present a modified stride length (commonly shorter). It is defined as:

$$S_k = x_r + x_l$$

where x_l and x_r is the step length of the left and right leg.

- *Range of motion.* Another value that can give an indication of a modified gait is the range of motion of the hip and knee joints in the sagittal plane (respectively HRoM and KRoM). It is calculated as the difference between the maximum and minimum flexion angle of the corresponding joint.

- *Peak flexion.* Similarly, the peak flexion angle of the hip (PHF, peak hip flexion) and knee (PKF, peak knee flexion) in the sagittal plane is also a good indicator or a modified gait pattern.

- *Foot clearance.* This variable is defined as the minimum distance between the foot and the ground at the MS. The foot drop problem (typically present in poststroke patients) usually causes a low foot clearance with a consequent risk of tripping. Hence, increasing the foot clearance is a positive outcome. This problem is typically addressed with an ankle support, however, it is possible to compensate it by increasing the knee and hip angles.

From the dynamic data it is possible to calculate the energy contribution of the exoskeleton as the ratio of powers (Λ) as defined in [25]:

$$\Lambda = \left| \frac{P_t - P_{ex}}{P_t} \right|$$

where P_t is the measured power of the user not wearing the exoskeleton, and P_{ex} is the power provided by the exoskeleton. $\Lambda = 1$ means that the prototype is not providing any assistance, while $\Lambda = 0$ represents 100% of assistance.

6.7.1.2 XoSoft Beta 1

Only one subject participated in the clinical testing of the XoSoft Beta 1 prototype. The participant was a 68-year-old male who experienced a cerebrovascular stroke 7 years before these trials. He presented unilateral gait impairment, with the right side affected. The Beta 1 prototype was configured to assist the hip and knee flexion on the right (affected) side, as described in Section 6.6.4. Three different configurations were tested:

- *Control strategy 0 (reference)*. In this configuration, the actuators were not mounted on the garment. This was used as the reference for the data analysis.
- *Control strategy 1*. In this case, the clutches were always active, meaning that the elastic bands were always affecting the subject, consequently behaving like a purely passive system. This configuration was useful to help understand the potential benefits of the clutch.
- *Control strategy 2*. This was the system working in normal conditions. The clutches were activated during midstance (around the 15% point of the gait cycle) and deactivated after 70 ms toe-off (around 70% of the gait). The timing was selected empirically from a preliminary testing of the system. It is important to mention that the hip and knee clutches were activated and deactivated simultaneously in order to reduce the number of control strategies. This is not the optimal control strategy, but a trade-off strategy decided after a clinical evaluation of the subject.

A number of different effects were observed during the testing. From the quantitative data, effectively the right PKF reached using control strategy 2 was 45 ± 13.8 degrees, against the 53 ± 3.2 degrees reached using the reference strategy. Also, the right HRoM was increased by14 degrees. These two factors indicate that the system has a positive effect on the subject's gait, potentially reducing the risk of tripping due to foot drop. With control strategy 1, the right PKF was increased even more (70 ± 3 degrees), but after a clinical analysis, it was considered that this was an overcompensation by the subject, without an evident benefit. In this case, the subject was trying to walk against the elastic straps that were always active. Regarding the step length asymmetry (SLA) and stride length, no representative differences were found with control strategy 2 with respect to the reference. More details about this analysis can be found in [27].

A direct analysis of the foot clearance also indicated a positive impact when using the XoSoft Beta 1 prototype. The foot clearance increased by 2−5 cm (depending on the condition), as described in [25]. An energy analysis was also done in order to understand how much XoSoft was assisting the subject. The average assistance provided by the XoSoft Beta 1 prototype was estimated to be $10.9\% \pm 2.2\%$ and $9.3\% \pm 3.5\%$ for the hip and knee, respectively.

From the qualitative point of view [28], the perceived wearer pressure was scored as 0 for all the body segments except for the assisted joints. In particular, the right lower leg has a value of 3 (moderate) while the right knee, upper leg, and hip were 2 (weak), indicating that the system did not produce any harming pressure on the subject. This means that from the comfort point of view, the system was acceptable. But also, that the level of assistance (estimated to be around 10% as described before) could be potentially increased within the limits of the maximum acceptable pressure.

The SUS questionnaire also gave promising results with a total score of 65, which can be translated as a usability between "ok" and "good." Being the first prototype of the system, this result was quite encouraging, but more importantly, the input from the user was fundamental for the development of the following prototypes.

Finally, the donning and doffing test gave also important feedback for the further improvement of the prototype, that is, the number of buckles and their position and the lateral zip were the main problems for the subject. Even if the subject was able to don and doff the system autonomously, the required time for the task was too high to be acceptable. It is important to consider that the subject had also a unilateral upper extremity impairment.

6.7.1.3 XoSoft Beta 2

Three further subjects, with ages between 55 and 79 years old, were included in the XoSoft Beta 2 laboratory testing. All of them had an incomplete SCI, but with a different modified gait pattern. Consequently, the actuator configuration and control strategy were also different for all of them. It is important to remember that while the XoSoft Beta 1 prototype was designed to assist knee and hip flexion on a single leg, the XoSoft Beta 2 prototype provided assistance for any configuration including ankle, knee, and hip in both legs. This was a very important improvement with respect to the Beta 1 prototype, which allowed testing of very different subjects.

Below, we list some of the positive and negative effects noticed with the Beta 2 prototype. It is important to indicate that each subject used a different actuation configuration, so the effects depend on each particular case. A full analysis of all the subjects is still a work in progress.

- *Whipping effect.* As described in the XoSoft Beta 1 analysis, on occasions the subject tried to overcompensate for the assistance of the exoskeleton. This caused an increase in the whipping effect (knee has overextension during the swing phase). The XoSoft Beta 2 prototype reduced this effect during normal walking, but it was still found to be present during slow walking.
- *Improved usability.* The same subject that tested the XoSoft Beta 1 prototype scored the XoSoft Beta 2 prototype with a 77.5 (high "good") for usability, an important improvement. One subject was more sensitive to skin pressure, due to his physical conditions, and another subject was not able to don and doff the system autonomously. This caused a lower usability score, with around 50 points, indicating a usability between "poor" and "ok."
- *Improved ground reaction impulse.* The GRF at the push-off was slightly increased with the XoSoft Beta 2 prototype. This increment improves the propulsion, reducing the energy required for walking. This effect was observed in several subjects, indicating a positive effect of the exoskeleton.
- *Donning/doffing.* Some of the subjects were not able to don and doff the system autonomously. This was identified as one of the limitations of the design and the motivation to follow a different design strategy for the garment.
- *Step length.* In one of the subjects, the step length was significantly increased, indicating also a potential benefit of the system.
- *Perceived pressure.* The perceived pressure was generally evaluated as low, except for the shoulders. This was due to the weight of the backpack (4 kg) containing the electronics and pneumatic components.

6.7.1.4 XoSoft Gamma

The main change perceived by the subjects between the XoSoft Beta 2 and XoSoft Gamma variants was the garment. Consequently, most of the findings of the XoSoft Gamma prototype were consistent with the Beta 2 prototype. However, a number of additional points were noted:

- *Perceived pressure.* The perceived pressure was in general similar to the Beta 2 prototype. Only one subject indicated a high pressure at the right lower leg. This was caused because of the system configuration for this subject, and a possible over stretch of the leg straps.
- *Usability.* The overall usability scores were similar with respect to the Beta 2 prototype. This indicates that the new design was going in a good direction, but still had room for improvement.
- *Kinematics.* One interesting effect that was found with one of the subjects was that while the system was assisting the hip flexion, the knee flexion was also increased during the swing phase. This effect was not noticed in this subject with the Beta 2 prototype because it was using a different configuration. This effect suggests that it is possible to provide assistance indirectly, that is, assist hip flexion to increase knee flexion. This will be further explored in the future.
- *Push-off.* While the plantarflexion actuation in the Beta 2 prototype was not very effective, with the Gamma prototype it was noticed that the subjects with this configuration had an increased support for the push-off.

6.7.2 Clinical testing

The clinical testing was carried out in a geriatric hospital (Geriatrie Zentrum Erlangen, Germany) with the aim of understanding the behavior and effects of XoSoft in realistic clinical conditions. A total of 11 subjects participated in the clinical testing, including five frail adults, one stroke patient, and five SCI patients. All these patients tested the XoSoft Beta 2 and XoSoft Gamma prototypes. The hospital did not have the required measurement equipment for a full gait analysis, as in the laboratory testing. Apart from a standard clinical gait analysis with an instrumented carpet, the evaluation was based on qualitative information, as described in the next subsection. The feedback provided by the subjects was very important for the development of the prototype, since it gave a very good indication of relative strengths and weaknesses of the system in real conditions.

6.7.2.1 Experimental protocol

The trials involved several tasks typically used in clinical evaluation, as well as specific tests for the evaluation of XoSoft:

- *Routine diagnostic instrumented gait analysis.* This test was carried out with the GATIRire sensorized carpet (CIR Systems Inc., Havertown, PA, United States) using the standard clinical procedure to evaluate patients.
- *Posturography* (static balance). The method includes four conditions: (1) normal standing with open eyes for 30 s, (2) normal standing with closed eyes for 30 s, (3) closed standing (feet together) with open eyes for 30 s, and (4) closed standing with closed eyes for 60 s.
- *Romberg test.* Closed standing with closed eyes for 60 s, as in the condition (4) of the posturography trial.
- *L-Test.* Starting from a seated position, the patient rises from the chair and walks for 3 m straight, turns 90 degrees to the right and then walks another 7 m straight. At that point, the patient stops and turns 180 degrees and does the inverse path until arriving back at the chair and sits down.
- *Figure-of-8-walk.* The patient, starting from a standing position, does a figure of eight around two cones.

The users were asked to provide feedback in different ways, as defined in [30]:

- *System usability.* For the evaluation of the system usability, the SUS forms used in the laboratory testing were used.
- *System perception.* A total of 16 questions were included in this questionnaire, and these were evaluated in the same way as SUS with a 5-point Likert scale. The questions are in five categories: Perceived Usefulness (PUS), Effort Expectancy (EE), Gerontechnology Self-Efficacy (SE), Anxiety (ANX), and Experiential Perception (EP). As per the SUS, this score goes from 0 to 100.
- *Perceived impact.* In a similar way, the perceived impact of the exoskeleton was evaluated through a questionnaire with 25 questions divided into nine categories: Attitude toward using the technology (ATUT), Anxiety (ANX), Gerontechnology Self-Efficacy (SE), Behavioral Intention (BI), Perceived adaptiveness (PAD), Social Influence (SI), Self-Liberty (SL), Quality-of-life Enhancement (QoLE), and Trust (TRUST). As per the SUS, this score goes from 0 to 100.

6.7.2.2 XoSoft Beta 2 versus XoSoft Gamma

The system usability was evaluated for both the XoSoft Beta 2 and Gamma prototypes, giving comparative results between those prototypes.

Table 6.6 SUS scores for 11 participants using XoSoft Beta 2 and Gamma prototypes.

SUS scores	Median	Min	Max
Beta 2	55.0	7.5	95.0
Gamma	62.5	35.0	97.5

The results, presented in Table 6.6, show an improvement in the usability of the system when moving from the Beta 2 to the Gamma variant. More details can be found in [31].

6.7.2.3 Exoscore XoSoft Beta 2

Exoscore comprises the three scores described before (system usability, system perception, and perceived impact). This new metric has been developed specifically to evaluate exoskeletons, and it is presented in [30]. It gives a good overview of the acceptance of the system, in a more detailed way than the SUS. Since this was the first usage of Exoscore, the main effort was put into understanding and implementing the tool. However, some preliminary results could be extracted from the Exoscore evaluation of the Beta 2 prototype (the detailed scores are presented in [30]).

From the total scores, the participants perceived the exoskeleton to be useful. Considering the age of the subjects, anxieties may become an important concern, and consequently the score of ANX should be a good indicator of the acceptance of the system. ANX scores were 76 in system perception and 70 in perceived impact, which indicates a good level of confidence. The EP construction, in the perception phase, indicates a lower result in the question regarding the appearance of the exoskeleton. This may indicate some concern from the subjects about the aesthetics of the exoskeleton, which is an important factor for the acceptance.

6.7.3 Home-Simulated Home environment validation

The final prototype (XoSoft Gamma) was tested in a simulated home environment. Performing the validation of the system in a simulated environment allowed us to monitor all the activities of the subjects and extract more relevant information. While a real home validation would be the ultimate goal, this was not currently possible due to safety and operational conditions.

The main objective was to evaluate the feasibility of the use of XoSoft in locomotion tasks related to daily life activities. Also, subjective feedback from the users was collected, using the same SUS questionnaires as in the laboratory and clinical environment.

The results of this validation are still under analysis. A preliminary evaluation has indicated that the acceptability and usability of the XoSoft Gamma prototype during the home-simulated testing was improved with respect to the laboratory and clinical testing. A more friendly environment and small improvements in the system configuration may be the main causes of an improved perception of the system.

6.8 Conclusions and future works

This chapter presented a full overview of the XoSoft development based on the UCD approach, from the gathering of information from the main stakeholders, to the testing and validation of the prototypes. Different user groups were involved during the entire development process, ensuring that the prototypes were evolving following real user needs. At the same time, novel technologies were explored in order to be able to integrate sensing and actuation in a complete soft wearable assistive device. This resulted in a modular and fully functional prototype that allowed testing in different conditions, from a laboratory environment to a clinical setting and ultimately to a home-simulated environment.

Testing and validation results showed a great potential for this kind of system, and these highlight the benefits of the UCD approach and soft technology, showing significant improvements between prototypes. This was especially critical since we were including completely new technologies and it was important to ensure a proper introduction of new elements between variants. The XoSoft Gamma prototype was considered quite satisfactory in clinical rehabilitation scenarios. The implementation of this device in home scenarios, where the user can wear the system during daily life activities, remains a challenge due to the current limitations of the system in terms of autonomy.

Several strengths were identified in the Gamma prototype. The modularity of the system and the garment design allow the usage of the system by very different subjects with various pathologies. While this is an advantage in rehabilitation scenarios, as highlighted before, it makes the system overcomplicated for home scenarios, due to the unnecessary number of configurations. A simplified version of XoSoft targeting a specific joint or a reduced number of joints would be the preferred path for a commercial device.

A critical point identified by most of the subjects was the weight of the system. While the level of assistance was in general well perceived by the subjects, the weight of the backpack (4 kg), containing electronics, battery, pneumatic system, etc., was an important limitation from the usability point of view. By improving this aspect, the overall acceptance of the system would be importantly increased. Since the pneumatic components represent an important part of the weight, the focus could be in the optimization of these components using a custom design. Another approach would be the replacement of the pneumatic actuation by another alternative technology (in the same way the electromagnetic clutches of the Beta 1 prototype were substituted by the soft pneumatic clutches of the Beta 2 and Gamma prototypes). Research into technologies for soft quasipassive actuation is and will continue to be a challenging and interesting topic for further development.

Regarding the actuation principle, it was demonstrated that the quasipassive actuation principles have a good potential to assist patients. However, we believe that there is still a margin for improvement, and it is still possible to increase the level of assistance in future generations of these devices.

The integration of the soft sensors was also very positive. Despite the technology being in an early stage of development, the integration with the garment offered a good performance in terms of knee angle estimation. Nevertheless, more work would be needed in order to improve the robustness of this solution.

References

[1] World Health Organization, Facts about ageing, <http://www.who.int/ageing/about/facts/en/>, 2014.
[2] P. Arnold, I. Bautman, The influence of strength training on muscle activation in elderly persons, Exp. Gerontol. 58 (2014) 58–68.
[3] M. Volkers, et al., Lower limb muscle strength: why sedentary life should never start, Arch. Gerontol. Geriatr. 54 (2012) 399–414.
[4] K. Strong, C. Mathers, R. Bonita, Preventing stroke: saving lives around the world, Lancet Neurol. 6 (2) (2007) 182–187.
[5] T. Yan, M. Cempini, C.M. Oddo, N. Vitiello, Review of assistive strategies in powered lower-limb orthoses and exoskeletons, Robot. Auton. Syst. 64 (2015) 120–136.
[6] S. Jezernik, G. Colombo, T. Keller, H. Frueh, M. Morari, Robotic orthosis lokomat: a rehabilitation and research tool, Neuromodulation: Technol. Neural Interface 6 (2) (2003) 108–115.
[7] J. Hidler, D. Nichols, M. Pelliccio, K. Brady, D.D. Campbell, J.H. Kahn, et al., Multicenter randomized clinical trial evaluating the effectiveness of the Lokomat in subacute stroke, Neurorehabil. Neural Repair. 23 (1) (2009) 5–13.

[8] J.F. Veneman, R. Kruidhof, E.E. Hekman, R. Ekkelenkamp, E.H. Van Asseldonk, H. Van Der Kooij, Design and evaluation of the LOPES exoskeleton robot for interactive gait rehabilitation, IEEE Trans. Neural Syst. Rehabil. Eng. 15 (3) (2007) 379–386.

[9] G. Zeilig, H. Weingarden, M. Zwecker, I. Dudkiewicz, A. Bloch, A. Esquenazi, Safety and tolerance of the ReWalk™ exoskeleton suit for ambulation by people with complete spinal cord injury: a pilot study, J. Spinal Cord. Med. 35 (2) (2012) 96–101.

[10] C.B. Baunsgaard, U.V. Nissen, A.K. Brust, A. Frotzler, C. Ribeill, Y.B. Kalke, et al., Gait training after spinal cord injury: safety, feasibility and gait function following 8 weeks of training with the exoskeletons from Ekso Bionics, Spinal Cord. 56 (2) (2018) 106.

[11] C. Tefertiller, K. Hays, J. Jones, A. Jayaraman, C. Hartigan, T. Bushnik, et al., Initial outcomes from a multicenter study utilizing the Indego powered exoskeleton in spinal cord injury, Top. Spinal Cord. Injury Rehabil. 24 (1) (2018) 78–85.

[12] C. Siviy, J. Bae, L. Baker, F. Porciuncula, T. Baker, T.D. Ellis, et al., Offline assistance optimization of a soft exosuit for augmenting ankle power of stroke survivors during walking, IEEE Robot. Autom. Lett. 5 (2) (2020) 828–835.

[13] F.L. Haufe, A.M. Kober, K. Schmidt, A. Sancho-Puchades, J.E. Duarte, P. Wolf, et al., User-driven walking assistance: first experimental results using the MyoSuit, 2019 IEEE 16th International Conference on Rehabilitation Robotics (ICORR), IEEE, June, 2019, pp. 944–949.

[14] J. Ortiz, E. Rocon, V. Power, A. de Eyto, L. O'Sullivan, M. Wirz, et al., XoSoft - a vision for a soft modular lower limb exoskeleton, Proceedings of the 2nd International Symposium on Wearable Robotics, WeRob 2016. Wearable Robotics: Challenges and Trends. Biosystems & Biorobotics, vol. 16, Springer, Cham, 2017, pp. 83–88. Available from: https://doi.org/10.1007/978-3-319-46532-6_14.

[15] J. Ortiz, C. Di Natali, D.G. Caldwell, XoSoft - iterative design of a modular soft lower limb exoskeleton, in: M. Carrozza, S. Micera, J. Pons (Eds.), Wearable Robotics: Challenges and Trends. WeRob 2018. Biosystems & Biorobotics, vol. 22, Springer, Cham, 2018, pp. 351–355. Available from: https://doi.org/10.1007/978-3-030-01887-0_67.

[16] E.B.N. Sanders, From user-centered to participatory design approaches, Des. Soc. Sci.: Mak. Connect. (2002) 1–8.

[17] V. Power, A. de Eyto, B. Hartigan, J. Ortiz, L. O'Sullivan, Application of a user-centered design approach to the development of XoSoft — a lower body soft exoskeleton, in: M. Carrozza, S. Micera, J. Pons (Eds.), Wearable Robotics: Challenges and Trends. WeRob 2018. Biosystems & Biorobotics, vol. 22, Springer, Cham, 2018, pp. 44–48. Available from: https://doi.org/10.1007/978-3-030-01887-0_9.

[18] International Organisation for Standardization, Ergonomics of human-system interaction — Part 11: Usability: Definitions and concepts, (ISO 9241-11 (2018) 2018 <https://www.iso.org/standard/63500.html>.

[19] V. Power, L. O'Sullivan, A. de Eyto, S. Schülein, C. Nikamp, C. Bauer, et al., Exploring user requirements for a lower body soft exoskeleton to assist mobility, Proceedings of the 9th ACM International Conference on PErvasive Technologies Related to Assistive Environments - PETRA '16 (1–6), ACM Press, New York, 2016. Available from: https://doi.org/10.1145/2910674.2935827.

[20] J. Ortiz, T. Poliero, G. Cairoli, E. Graf, D.G. Caldwell, Energy efficiency analysis and design optimization of an actuation system in a soft modular lower limb exoskeleton, IEEE Robot. Autom. Lett. 3 (2018) 1. Available from: 10.1109/LRA.2017.2768119.

[21] J.R. Amend, E. Brown, N. Rodenberg, H.M. Jaeger, H. Lipson, A positive pressure universal gripper based on the jamming of granular material, IEEE Trans. Robot. 28 (2) (2012) 341–350.

[22] A. Sadeghi, A. Mondini, B. Mazzolai, Preliminary experimental study on variable stiffness structures based on textile jamming for wearable robotics, in: M. Carrozza, S. Micera, J. Pons (Eds.), Wearable Robotics: Challenges and Trends. WeRob 2018. Bio-systems & Biorobotics, vol. 22, Springer, Cham, 2018, pp. 53−57. p. 49-52. https://doi.org/10.1007/978-3-030-01887-0_10.

[23] A.F. Hidalgo, J.S. Lora-Millán, E. Rocon, IMU-based knee angle estimation using an extended kalman filter, 2019 41st Annual International Conference of the IEEE Engineering in Medicine and Biology Society (EMBC), IEEE, July 2019, pp. 570−573.

[24] M. Totaro, T. Poliero, A. Mondini, C. Lucarotti, G. Cairoli, J. Ortiz, et al., Soft smart garments for lower limb joint position analysis, Sensors 17 (2017) 1. Available from: 10.3390/s17102314.

[25] C. Di Natali, T. Poliero, M. Sposito, E. Graf, et al., Design and evaluation of a soft assistive lower limb exoskeleton, Robotica (2019) 1−21. Available from: 10.1017/S0263574719000067.

[26] C. Di Natali, A. Sadeghi, A. Mondini, E. Bottenberg, B. Hartigan, B. De Eyto, et al., Pneumatic Quasi-Passive Actuation For Soft Assistive Lower Limbs Exoskeleton. Frontiers in Neurorobotics, special issue on Application of Bio-Inspired Concepts in Gait Assistance, Front. Neurorobot., 30 June 2020 | https://doi.org/10.3389/fnbot.2020.00031.

[27] M. Sposito, T. Poliero, C. Di Natali, J. Ortiz, C. Pauli, E. Graf, et al., Evaluation of XoSoft Beta-1 lower limb exoskeleton on a post stroke patient, in: Sixth National Congress of Bioengineering, Milan, Italy, June 25−27, 2018.

[28] S. Eveline, E.S. Graf, C.M. Bauer, V. Power, A. de Eyto, E. Bottenberg, et al., Basic functionality of a prototype wearable assistive soft exoskeleton for people with gait impairments a case study, PETRA '18 Proceedings of the 11th PErvasive Technologies Related to Assistive Environments Conference (202−207), ACM Press, New York, 2018. Available from: https://doi.org/10.21256/zhaw-3876.

[29] G. Borg, Borg's Perceived Exertion and Pain Scales, Human Kinetics, 1998.

[30] L. Shore, V. Power, B. Hartigan, S. Schülein, E. Graf, A. de Eyto, et al., Exoscore: a design tool to evaluate factors associated with technology acceptance of soft lower limb exosuits by older adults, Human Factors (2019). 0018720819868122.

[31] E. Graf, C. Bauer, S. Schülein, A. de Eyto, V. Power, E. Bottenberg, et al., Assessing usability of a prototype soft exoskeleton by involving people with gait impairments, in: WCPT World Confederation for Physical Therapy Congress, ZHAW Zürcher Hochschule für Angewandte Wissenschaften, Geneva, May 10−13, 2019.

CHAPTER 7

TwAS: treadmill with adjustable surface stiffness

Amir Jafari and Nafiseh Ebrahimi
Advanced Robotic Manipulators ARM Lab, Department of Mechanical Engineering, College of Engineering, University of Texas at San Antonio UTSA, San Antonio, TX, United States

7.1 Introduction

During the course of rehabilitation for mobility impaired patients, accurately adjusting the walking/running conditions based on the current status of the patients as well as the set goals of rehabilitation are essential for achieving the optimal output. Of those locomotion conditions, surface stiffness is an important factor as it affects both the kinematics and dynamics of locomotion. Usually walking on softer surfaces requires more muscle activation as more force would be required for moving forward and will happen at lower pace as the balance might be threatened by the surface compliance. The main idea of rehabilitation, however, is to start simple and then gradually progress to harder efforts.

So far in the rehabilitation prescriptions for these patients, one can usually notice walking on compliant surfaces. However, the available apparatus for this purpose in rehabilitation sections is limited to foam walkways. The problems with these types of walkways are (1) their limited lengths and (2) their fixed and limited surface stiffness levels. It is important to consider that rehabilitation is usually an intense process if optimal output is intended, especially right after the cause, such as stroke. Limited lengths of these walkways mean that the patient would have to walk on the foam from one end to the other, of course with the help of the therapist, and then again walk back to the initial end. That makes the rehabilitation process very time-consuming which not only negatively affects the output performance of the patient but also the financial burden on the patients, their families, care givers, and the government health system.

In addition, if the level of surface stiffness is not suitable for the user, walking on it not only would not assist the optimal outcome but also

Soft Robotics in Rehabilitation
DOI: https://doi.org/10.1016/B978-0-12-818538-4.00007-1

might lead to more serious problems and injuries. In fact, injuries at the ankle joint are a very common type of injury for athletes who play sand types of sports activities. Therefore the limited surface stiffness of the currently available foam walkways is not only a possible source of risk but also many times it may lead to not including foam walking in the rehabilitation process.

A patient's weight is also another important factor that can alter the rehabilitation process. Many poststroke patients are not able to walk because they cannot support their own weight during the walking activity, especially during initial stage poststroke. These patients usually are held by the therapist during the action which, again considering the desired high intensity of the rehabilitation process, is indeed a limiting factor. Harness systems usually are used to help these patients to support their own weight for walking on different walkways in rehabilitation sections, which, again considering the limited lengths of the available walkways, does not lead to an optimal outcome.

Another important factor in achieving the optimal outcome is providing accurate and reliable data on the performance of the patient to the therapist during the course of rehabilitation. So far, these types of data are visually based, meaning that the therapist would just simply watch the performance of the patient and judge whether or not it is in line with what is expected or not. This method of analysis has two implications: (1) it is based on the experience of the therapist and as a result it may be different from one therapist to another for a single patient; (2) only the kinematics of walking can be captured and probably some verbal feedback from the patients on how they feel, if there is pain in any joint or if they feel comfortable with the walking speed and any other physical condition of the walking environment such as surface stiffness. If we can provide accurate and numeric data to the therapist we can avoid experience-based evaluation and so the data on the performance of the patient can be easily transferrable from one therapist to another. Also, with real-time accurate data, a therapist can then more reliably judge the performance of the patient at any given time during the course of rehabilitating and accordingly decide on the next steps.

In this chapter, we introduce a novel dual belt Treadmill with Adjustable Stiffness (TwAS), by which one can adjust the level of surface stiffness and speed of each belt, accurately. TwAS is integrated with the following systems: a weight cancelation hydraulically actuated harness system LiteGait, motion capture system with eight Vicon cameras, and a

ParvoMedics VO_2max metabolic cost measurement system. With this integrated system, a mobility impaired patient can walk at different speeds, on different surface stiffness levels, while the walking kinematics can be tracked by the motion capture system and the energy expenditure can be measured by the VO_2max system. Furthermore, using the model-based surface stiffness level together with the tracked surface deflection using the motion capture system, the ground reaction force can be calculated, which can be used to understand the joints' torques through inverse dynamic approaches. This system can provide a unique rehabilitation and evaluation platform for therapist as it can provide real-time data on the performance of mobility impaired patients.

The stiffness adjustment mechanism of TwAS is based on that of Actuator with Adjustable Stiffness (AwAS). In this chapter, we first explain the stiffness adjustment mechanism of AwAS, its properties, such as the energy required to change the stiffness, the decoupling between force and stiffness, the force required to maintain the level of stiffness, the ratio between the energy storage and release, etc. The stiffness adjustment mechanism of TwAS will be explained in detail together with a discussion on the material selection. Finally, some preliminary results regarding the effect of surface stiffness on the walking gait and energy expenditure will be presented.

7.2 Actuator with Adjustable Stiffness mechanism

This section reviews the relevant literature in variable stiffness mechanisms. Depending on how elastic elements and motors are used in these mechanisms, in general they can be classified into two main groups: antagonistic and series configurations. In the following, these two groups are explained in detail.

7.2.1 Antagonistic configurations

In antagonistic setups, two motors are antagonistically actuating a joint while each one is connected to a nonlinear elastic element [1]. Usually corotation of the motors will result in changing of the position while tuning the stiffness is achieved by counterrotation of the motors. In some cases, one motor changes the position and the second motor change the pretension of both elastic elements [2]. This type of configuration has been realized in three different formats: simple antagonistic, bidirectional antagonistic, and cross coupling. In the following each format is presented and some well-known corresponded antagonistic actuators are explained in detail.

7.2.1.1 Simple antagonistic

This setup (Fig. 7.1), as the name says, is the simplest format in the family of antagonistic setups. This configuration is actually mimicking animal and human joints in which two muscles (agonistic and antagonistic) can move a limb and adjust its compliance, Fig. 7.2. The reason that the stiffness of antagonistic springs is shown by different letter K_1 and K_2 is to highlight that an antagonistic configuration is not necessarily a symmetrical configuration [3,4]. Even different actuating motors can be used in these antagonistic setups. Muscles only can pull and are not able to push the joint. Compliance is an inherent property of muscles with a nonlinear force to extension profile. This makes the joint stiffer as the joint's deflection increases. Therefore they can be modeled as unidirectional compliant actuators.

7.2.1.1.1 Biological inspired joint stiffness control

Petit et al. [5] described a device based on the antagonistic setup of two nonlinear springs. This biological inspired joint stiffness control is a rotational joint, actuated by two series elastic actuators, as shown in Fig. 2.3. This figure shows how the linear characteristics of the spring is transferred into a quadratic characteristic, by using special shaped pieces, over which two wheels roll. The centers of the wheels are interconnected by a linear spring. When both servomotors rotate in the same direction, the equilibrium position of the joint is changed. When they rotate in the opposite direction, the stiffness of the joint will be changed. Fig. 7.3 (left) shows a mechanical schematic of the joint used in their research. α and β (Fig. 7.3, right) are the agonist and antagonist servo angles, and θ is the joint angle. As a convention, positive rotation is clockwise for α and

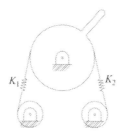

Figure 7.1 Simple antagonistic configuration.

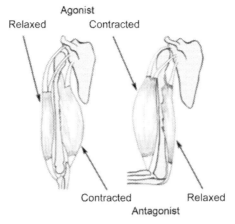

Figure 7.2 Agonistic and antagonistic muscles in human joint.

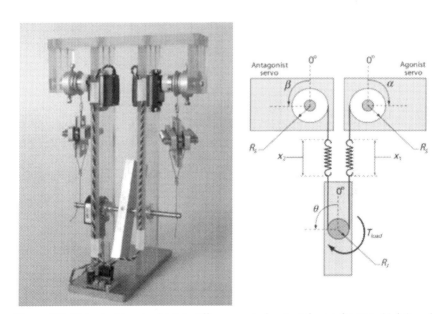

Figure 7.3 Biological inspired joint stiffness control: principle mechanism (right) and actual prototype (left).

counterclockwise for β and θ [3]. To avoid cable slack, they define the operating region of the joint to be $\theta \in \{-\beta \leq \theta \leq \alpha\}$. Although linear springs are easier to work with and are readily available, they do not allow the joint stiffness to be varied. The stiffness that results from their use is a function of both stiffness and radius of the pulley.

7.2.1.1.2 Actuator with mechanically adjustable series compliance

Actuator with mechanically adjustable series compliance (AMASC) [6], as shown in Fig. 7.4, was developed specifically for the purpose of actuating a running robot. It actuates a single degree of freedom, with a large spring in series between the electric motor and the output of the device. The spring is sized to store the energy of a running gait, so the robot may bounce on the spring much like a rider on a pogo stick, or like any animal in a regular running gait. The spring stiffness is mechanically adjustable; therefore, leg stiffness can be tuned for a particular gait or ground surface.

The AMASC is an integrated mechanism and software controller, with mechanical design choices made to closely match a simple mechanical model. The software controller is based on the same simple mechanical model, safely ignoring most of the complexities of the actual mechanism. This model is illustrated in Fig. 7.5, in two different forms: one rotational, one linear. The rotational model is physically similar to the prototype AMASC, while the linear model is a simpler form that still captures the important properties of the system.

In both models, the dynamics of the system controlling the pretension, x_3, are ignored. The pretension is entirely unrepresented in the linear model, and the spring stiffness, K_e, is assumed to simply be a programmable value. The AMASC is essentially a single compliant joint, most closely resembling a knee, endowed with engineered natural dynamics. There are two degrees of freedom, and two corresponding motors. One motor controls the spring pretension. As shown in Fig. 7.5A, there are two identical opposing springs, much like antagonistic muscles in animals. The pretension, x_3, stretches both springs, which is analogous to muscle cocontraction in animals. The knee joint does not move, but its rotational stiffness

Figure 7.4 Actuator with mechanically adjustable series compliance.

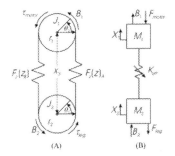

Figure 7.5 Simple linear and rotational model of AMASC. (A) Rotational model; (B) linear model.

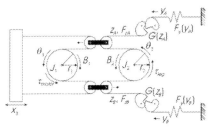

Figure 7.6 Cable routing diagram of the AMASC. J_1 and J_2 are pinned in place but can rotate freely.

increases with the pretension, allowing the actuator to tune the stiffness aspect of its natural dynamics. The other motor controls the spring rest position (θ_1 in Fig. 7.5A and x_1 in Fig. 7.5B), which is used as the primary energy source and controls any motions not described by the system's natural dynamics. These two motor-controlled parameters, along with the leg angle at touchdown, are the parameters necessary to control SLIP model running.

In order to create a low-friction, zero-backlash system, the AMASC utilizes a high-speed cable drive [7]. There is some stretch in the cable transmission, which adds series compliance to the system, and is incorporated into the effective spring constant of the model. Because the cables are round, they may wrap around pulleys placed at any angle, unlike standard belts. This design freedom makes it easier to route cables through joints, allowing the motors to be located remotely. Fig. 7.6 shows the cable routing, illustrating the role of each motor in the tension of the two springs. Also shown is the fact that a displacement of the leg (θ_2 or x_2) results in displacement of the motor (θ_1 or x_1), displacement of the springs, or some combination of the two.

There is a speed reduction between the first and second pulleys not shown on the diagram; it is implemented using a combination of a block-and-tackle pulley mechanism and a difference in radii between r_1 and r_2. The speed reduction is physically located near the knee joint, but diagrammatically located near the motor, θ_1. In all of our representations, the speed reduction is shown solely as a difference between r_1 and r_2: All friction related to the speed reduction is applied to θ_1 and corresponds to B_1. The inertia of the speed reduction is added to the inertia of the motor, and corresponds to J_1. A speed reducer amplifies the motor inertia by the square of the speed reduction; this amplification appears in the relatively large values of J_1. The transmission between θ_2 and the springs has very low friction, and no speed reduction. Because the high-frequency behavior of the system is generally handled by the springs, low friction and inertia are most important in this part of the AMASC. The low-frequency behaviors of the system are handled by the motor, and thus friction and inertia can be overcome by relatively low-bandwidth software compensation. Perhaps the most important aspect of the AMASC, along with the series spring, remotely located motors, and zero-backlash cable drive, is the physically variable series compliance. As was mentioned in Chapter 1, High-Performance Soft Wearable Robots for Human Augmentation and Gait Rehabilitation, physical compliance is crucial for a running gait, while varying the compliance is a useful control strategy. The AMASC's physical compliance resides in unidirectional fiberglass plates, which have a relatively high work capacity on the order of 1000 J/kg. Varying the compliance of the AMASC is achieved in much the same way as in animals, with cocontraction of opposing nonlinear springs which again needs a high amount of energy consumption.

As can be seen in Fig. 7.7, the AMASC spring function becomes stiffer at increasing levels of pretension. The spring function is not exactly linear in defection, although this can be remedied through pulley design. It is also apparent in Fig. 2.8 that there is some hysteresis due to mechanical friction.

7.2.1.1.3 Pleated pneumatic artificial muscle

The pleated pneumatic artificial muscle (PPAM) [8] consists of a membrane made of an aromatic polyamide (Fig. 7.8), such as Kevlar, to which a thin liner of polypropylene is attached in order to make the membrane airtight. The membrane of this muscle is arranged into radially laid out

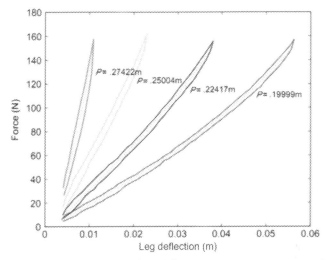

Figure 7.7 Spring force response at four different pretension settings, with one possible curve fit (dotted line).

Figure 7.8 Pleated pneumatic artificial muscle.

folds that can unfurl free of radial stress when inflated. Reinforcing high tensile Kevlar fibers are positioned in each crease.

The high tensile longitudinal fibers of the membrane transfer tension, while the folded structure allows the muscle to expand radially, which avoids energy losses and hysteresis. The folded membrane is positioned into two end fittings, which close the muscle and provide tubing to inflate and deflate the enclosed volume. An epoxy resin fixes the membrane and the Kevlar fibers to the end fittings. Van Ham et al. (2002) extensively discuss several characteristics concerning the PPAM, for inelastic as well as elastic membranes. Mainly two characteristics are important: generated traction and enclosed volume for each contraction. The first is used for joint torque dimensioning and control purposes, while the latter is used to predict joint compliance with closed muscles. Furthermore, the maximum diameter when the muscle is fully bulged should be taken into account when designing the different joints of the robot in order to provide enough space for the muscle to bulge. When inflated with pressurized air, the muscle generates a unidirectional pulling force along the longitudinal axis, as is shown in Fig. 2.10 for different levels of pressure.

7.2.1.2 Cross coupling antagonistic

In simple antagonistic setup when maximum stiffness is achieved both springs are fully deflected (either extended or compressed depends on the spring's type), therefore there would be no torque anymore at the joint to rotate the link [2,9,10]. This problem is solved by adding another spring which couples two motors as it is shown in Fig. 7.9. This configuration is known as cross coupling.

Another role of this spring is to shift the range of stiffness to a desired working range. In the simple configuration, when both motors are at their rest position, the minimum stiffness is achieved and both springs face a minimal deflection (the minimal deflection cannot be zero because again no force can be delivered to the joint to move it). However, in cross coupling setups, when both actuators are at their zero position, stiffness of the

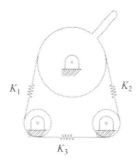

Figure 7.9 Cross coupling antagonistic configuration.

joint is not minimum since springs are under tension by the third spring and are not at their minimal deflection. To increase the stiffness both motors have to deflect more the two springs and as a result relax more the third one, while to decrease the stiffness both motors have to relax more the two springs and deflect more the third one. The third spring may or may not be at the same type of the other two springs.

7.2.1.3 Bidirectional antagonistic

In the robotic world, further progress to this mammalian configuration has been incorporated to improve the efficiency and performance of robotic joints. For instance, since each motor in the basic configuration can actuate the joint only in one direction, either push or pull, therefore only half of each motor power can contribute to the joint torque [9,11]. To improve the efficiency, as shown in Fig. 7.10, two additional nonlinear springs, but with opposite force directions to the other two, are connected to motors. In this case, each motor can simultaneously pull and push the joint. This configuration is known as the bidirectional antagonistic setup.

7.2.2 Series configurations

In series setups, two motors are located in series. One motor is assigned to set the position, the other one works on the elastic element to adapt the stiffness. The most common way to tune the stiffness is changing pretension of the springs [12,13].

7.2.2.1 Series configurations based on the pretension

In this type of series setups, as shown in Fig. 7.11, one motor changes the position of the link. Between this motor and the link there is an elastic element (either linear or nonlinear, however in case of linear elastic

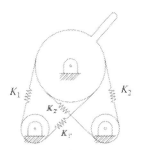

Figure 7.10 Bidirectional antagonistic configuration.

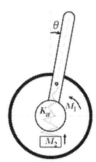

Figure 7.11 Series configuration: motor M_1 changes the position while motor M_2 changes the pretension of the elastic element to regulate the stiffness.

elements, there has to be a mechanism which generates a nonlinear profile of stiffness vs deflection). The second motor then changes the pretension of the elastic elements in order to regulate the stiffness [14–19].

7.2.3 Drawbacks

Here we explain the drawbacks of all the previously mentioned configurations (and the corresponding actuators) in order to highlight the motivation of the current research.

In all of these configurations (either antagonistic or series), regulating the stiffness is done through changing the pretension of elastic elements. This means that:

1. To change the stiffness, considerably large force and thus energy is needed. There is always a force because the elastic element is deflected. Therefore to tune the stiffness this force has to be overcome [20].
2. In the case of antagonistic configurations, to regulate the stiffness two motors have to contract each other (also they need to contract the force of the elastic element's deflection). Therefore a lot of energy is wasted so the energy efficiency in these types of variable stiffness actuators is very low [21].
3. Position of the link and stiffness are not decoupled. Stiffness depends on the deflection of the elastic elements. If the link deviates from its equilibrium position, then it affects the deflection of the elastic elements and therefore stiffness will also change [22].
4. As a consequence of drawback 3, to maintain the stiffness while the position is changing, considerable energy is needed in order to change

the pretension to compensate for deflection variation due to changing the position.
5. As a consequence of drawback 1, strong motors are needed, which results in a heavy and big setup.
6. Energy efficiency of these actuators is low. Only a very small part of the energy which is required to regulate the stiffness can be stored in the elastic elements and return to the link.

7.2.4 Actuator with Adjustable Stiffness configuration

The AwAS configuration is a kind of series configuration [23–25]. An essential element in this configuration is the lever. The lever has three critical points. One point is the point at which the lever can rotate. This point is called the pivot. Another point is that at which the force is applied. The third point is the connection point between the lever and elastic element. In this configuration, to regulate the stiffness, the pretension of the elastic element is kept fixed, but instead the relative position of these three points at the lever changes. Fig. 7.12 shows an example of this configuration in which the stiffness is tuned by changing the position of elastic elements and the pivot of the lever is aligned with the center of rotation of the link and position motor (the motor which changes the position). The proposed configuration has several advantages over other types of variable stiffness configurations, among which energy efficiency is the most essential one.

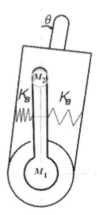

Figure 7.12 The AwAS configuration: motor M_1 changes the position while motor M_2 changes the position of the elastic element to regulate the stiffness.

7.2.4.1 Actuator with Adjustable Stiffness mechanism

Here we explain the design and development of the first prototype (Fig. 7.13) of AwAS which is based on the proposed configuration.

The load is applied on the other end of the lever. Two linear springs are located antagonistically on both sides of the lever between the pivot and the point at which force is applied. The distance between the pivot and springs can be defined as the lever arm. Stiffness of the link depends on the lever arm. The larger is the lever arm, the stiffer is the link. Fig. 7.14 shows this concept in two situations: stiff (left) and soft (right). In the following the electromechanical design of the AwAS will be explained in detail.

In AwAS like any other variable stiffness actuators, two motors are used: one to change the position and the other one to tune the stiffness. Since AwAS is energy efficient variable stiffness actuator, then to regulate the stiffness it does not need a big and strong motor compared to

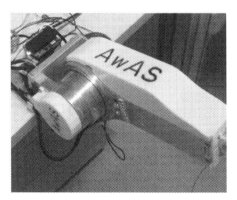

Figure 7.13 Physical covered setup of AwAS.

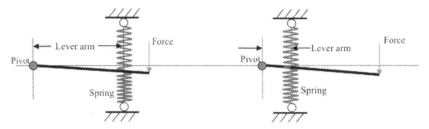

Figure 7.14 Schematic of the proposed configuration: lever rotates around the pivot, force is applied to one end of the lever. Springs are placed between the load and the pivot. The distance between the springs and the pivot is the lever arm.

the position motor (the motor which changes the position). The position motor (hereafter called M_1) is connected to a gearbox. The output of the gear box is then connected to a link which is called the intermediate link. Two linear springs are antagonistically placed on both sides of the intermediate link. The distance between the spring and center of rotation of M_1 is the lever arm. The second motor which is the stiffness motor (hereafter called M_2) is then assembled on the intermediate link and through a mechanism which converts rotary motion of M_2 into a linear motion changes the position of the springs and therefore changes the lever arm.

As was mentioned before, springs are connected to the intermediate link from one side. From the other side, they are connected to another link which is called the output link (the output link actually plays the role of lever which rotates around the center of rotation of motor M_1, so center of rotation of M_1 is the pivot point of the lever). Therefore between the intermediate link (which is connected to the output of the motor M_1) and the output link, springs are generating the compliance of the system. The guiding mechanism driven by another motor M_2 allows the control of the length of the arm by moving the two springs toward (to reduce stiffness) and away from (to increase stiffness) the center of rotation.

The sum of the lengths of the two springs is always a constant, so the pretension does not change when controlling the stiffness. When the output link is in its equilibrium position (the angular position where zero torque is generated, so that extension of both springs is equal), then the force generated by the springs is perpendicular to the displacement needed to change the stiffness. This has the important consequence that in principle no energy is needed to change the stiffness. In different designs the force is always parallel to the displacement requiring a strong motor and sufficient amount of energy to change the stiffness. In reality, the presence of friction has to be overcome. In addition, if the joint is not in the equilibrium position the force generated by the spring has a small component parallel to the displacement and a small amount of energy is needed. However due to this property the motor controlling the stiffness can be significantly smaller than that in other designs of variable stiffness actuators. An additional advantage of this design is that it does not require the use of nonlinear springs or mechanisms to provide the nonlinear force/displacement profile which is necessary for the stiffness regulation.

7.2.4.2 Second version of Actuator with Adjustable Stiffness mechanism

Here we explain the design and development of the second version of AwAS (AwAS-II) which is similar to the proposed configuration as it employs the lever to regulate the stiffness. However, in AwAS-II, the position of the springs is fixed. A prototype of this actuator is shown in Fig. 7.15.

As was mentioned before, the proposed configuration is based on a lever mechanism. In the lever, there are three important elements: pivot point (the point around which the lever can rotate); spring point (the point at which springs are placed); and force point (the point at which the force is applied). In AwAS, to regulate the stiffness, the pivot point was aligned with the center of rotation of the position motor and the second motor M_2 changes the spring point. However, in the lever mechanism applied to AwAS-II, the spring and force point is kept fixed but instead the pivot point is moving. As shown in the Fig. 7.16 the lever can rotate around its pivot which is located between the spring and force.

When the pivot is close to the spring point the lever is soft and when it is near to the force point the lever becomes stiff. The distance between the pivot and spring point is called L_1 and the distance between the pivot and force point is called L_s. Stiffness of the lever depends on the ratio L_1/L_2. This ratio is zero when the pivot is aligned with the spring point, in this case the lever stiffness is zero which means that the lever can easily rotate around the pivot without any resistance due to springs deflection (springs in this case are not deflected). By moving the pivot away from springs the ratio and thus the stiffness increase continuously. When the pivot reaches the force point, the ratio and thus the stiffness will become infinite. In this case the lever is completely locked and cannot rotate

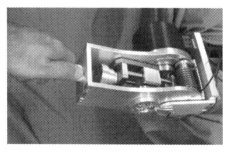

Figure 7.15 Physical setup of AwAS-II.

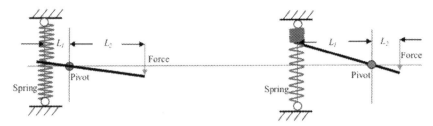

Figure 7.16 Schematic of the proposed configuration applied to AwAS-II: force is applied to one end of the lever while springs are placed at the other end. The lever rotates around the pivot which is between force point and spring point.

around the pivot. Therefore the range of stiffness is from zero to infinite and this range does not depend on the lever length and spring stiffness (in AwAS the stiffness range depends on these parameters). This has a very important advantage: a short lever and a pair of soft springs can be used, which help to make a lighter and more compact setup compared to other variable stiffness setups (including AwAS). Since AwAS-II is also based on the lever mechanism, stiffness regulation is energetically efficient.

7.2.5 Experimental results on Actuator with Adjustable Stiffness adjustment mechanism

Here we explain the experiments which have been conducted for AwAS and AwAS-II. Since these are variable stiffness actuators, first their ability to tune the stiffness is studied. Energy consumption to regulate the stiffness is a very important aspect of these actuators. We claimed that these actuators (and in general those based on the lever principle) are energy efficient actuators. This will be proved through a set of experiments which have been done on AwAS and will be explained here.

7.2.5.1 Preliminary experiments

As preliminary experiments, the capability of AwAS and AwAS-II to regulate the stiffness, stiffness response, tracking of a trajectory (both for position and stiffness), and the required energy to change the stiffness for AwAS are studied.

First the ability of AwAS and AwAS-II to regulate the stiffness are studied. The position of the intermediate link (position of the motor M_1) is fixed. Then the output link is manually pulled to deviate from its equilibrium position. The difference between readings of the two encoders (one for the position of the intermediate link and the other one for the position of the output link) gives the angular deflection. The torque is measured by the torque sensor (strain gage). Fig. 7.17 shows the torque versus angular deflection for different arms which is experimentally plotted for AwAS. In this figure, the increased slope of the curves as the lever arm becomes longer reveals the effect of the stiffness increase. Fig. 7.18 shows this plot for AwAS-II for different ratios. Here again the slope of each curve represents the corresponding stiffness for certain values of the ratio. As is clear from these two graphs, while AwAS can achieve a relatively large range of stiffness regulation, AwAS-II regulates the stiffness in the largest possible range from zero to nearly infinite. It should be notified here that infinite stiffness cannot be achieved because in any case the level of stiffness cannot exceed the structural stiffness of the joint.

Fig. 7.19 introduces experimental stiffness change step responses for AwAS from 64 to 250 Nm/rad and from 250 to 1024 Nm/rad. It can be observed that the capability of the stiffness regulation drive to tune the stiffness to the desired amount with good fidelity. Although step inputs commands are shown in Fig. 7.19, the controller of the motor M_2 was fed with a reference trajectory generated by a minimum jerk trajectory module. The stiffness in both cases is estimated based on the value of the arm.

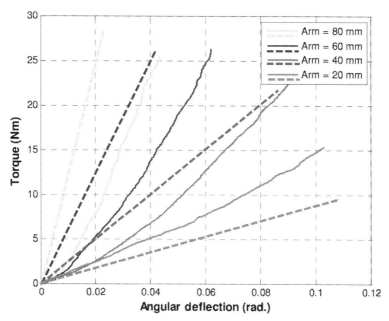

Figure 7.17 Torques curves versus angular deflection of AwAS for different arms.

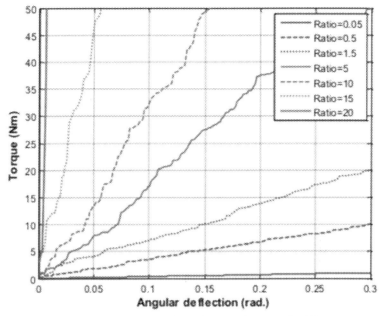

Figure 7.18 Torques curves versus angular deflection of AwAS-II for different ratios of α.

Both motors M_1 for position and M_2 for stiffness were simultaneously controlled to follow the sinusoidal position and stiffness trajectories of different frequencies. Fig. 7.20 presents the output link position and stiffness trajectories against the reference ones revealing the capability of the actuator to control both variables independently with good fidelity.

We then fixed the position of the intermediate link. Then to achieve different angular deflection, the position of the output link is also fixed to different values. For each case stiffness is set to the minimum and then increased to its maximum. To calculate the required energy to change the stiffness, the current supplied to the motor M_2 is measured through a current sensor placed between the power supply and motor M_2. Fig. 7.21 shows this required energy. It is important to note here that the angular deflection is from −0.2 to +0.2 rad. However, the full range of angular deflection cannot be achieved for all the levels of stiffness. When the stiffness is high (when the arm is large) before reaching the maximum angular deflection, the springs become fully deflected, which prevents the angular deflection from increasing more. The expected energy consumption surface profile for AwAS demonstrates a good resemblance with the experimentally obtained data plotted in Fig. 7.21. The higher energy values

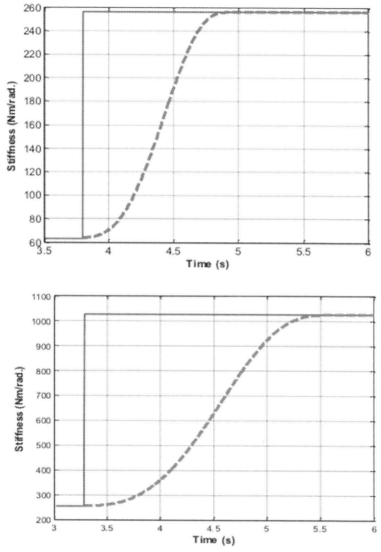

Figure 7.19 Stiffness response from 64 to 250 Nm/rad (top) and from 250 to 1024 Nm/rad (bottom).

depicted in the experimental graph are due to the fact that the theoretical energy consumption surface shown in Fig. 7.21 is based on the mechanical work done without taking into account any dissipative components such friction and damping, or energy losses when converting electrical energy into mechanical energy.

TwAS: treadmill with adjustable surface stiffness 219

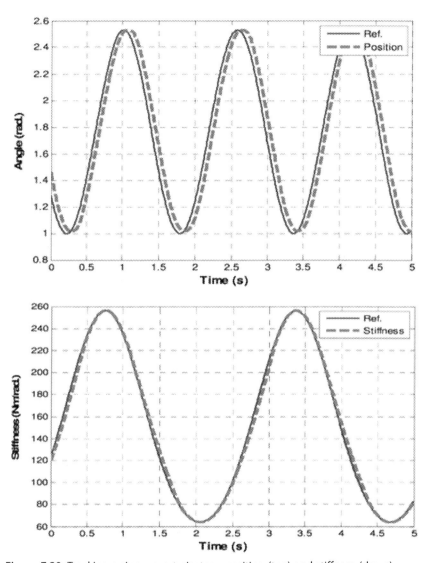

Figure 7.20 Tracking a sine wave trajectory; position (top) and stiffness (down).

7.2.6 Treadmill with Adjustable Stiffness

Based on the stiffness adjustment mechanism in AwAS-II, we implemented a stiffness adjustment mechanism in a treadmill system. The reason we were interested in making a treadmill with adjustable surface stiffness is its application in walking rehabilitations.

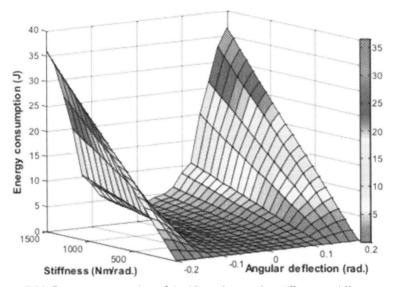

Figure 7.21 Energy consumption of AwAS to change the stiffness at different angular deflections based on the current delivered to M_2.

Bipedal movement is an extremely difficult engineering and controls problem. Scientists and engineers are tasked with the goal of emulating this motion. The purpose of this research and development is to better understand how human muscles and joints operate in conjunction with one another. By understanding muscle energy consumption and cohesion, scientists are better equipped to handle the further research of the human gait cycle.

In order to better understand the human gait cycle and other bipedal movement, researchers have developed exoskeletal systems that can be used in conjunction with various data acquisition systems to gather information on the forces and frequencies experienced. The main problem occurs when trying to simulate real-world conditions for the test subject to walk, jog, and run in. For instance, it is quite noticeable that a person will have to exert more energy to run on sand than on concrete. The reason for this is that different surfaces and terrains have different surface stiffness and compliances. To replicate the various surface stiffnesses that a person would experience in the real world is expensive for a research lab to accomplish. Also, going to the site for the experiment would be costly and time-consuming.

7.2.6.1 Performance specification

Load: The maximum load that the TwAS will have to support is 450 lbs. This includes the weight of the operator as well as the exoskeleton suit.

Speed: The TwAS has to vary between speeds that require the test subject to walk, jog and run on the selected treadmill. These speeds range from 0 to 3.9 mph, 4 to 5.9 mph, and 6 to 8 mph for walking, jogging and running, respectively.

Variable stiffness: The overall effective stiffness of the system has to be able to change from virtually zero to virtually infinite. This stiffness needs to cover this range within 5 s of operation.

7.2.6.2 Design specification

Overall dimensions: The TwAS must fit within a max floor space of 132 × 108 in. The height of the design should not exceed 24 in.

Power requirements: Usually we have a standard NEMA 5−15 R duplex outlet (grounded outlet) that is rated for 15 A and 120 V. The treadmill selected must operate under this power.

Vertical compliance: A maximum of 6 in. (15.24 cm) vertical compliance will be permitted during the operation of the TwAS. The vertical direction will be the only degree of freedom of the treadmill platform while the test subject is on the TwAS.

This treadmill requires 1.5 hp to operate. The operating speed is from 1 to 8 mph. Belt size is 14 × 38.5 in. The power consumption is 110 V with 500 W. Overall treadmill dimensions are approximately 49 × 24 × 11 in. Finally, the weight capacity of this model is 250 lbs. (113.4 kg).

By purchasing a prebuilt treadmill, Vulcan will have an easier time working with the interfacing of controls and motors. This treadmill will be disassembled to fit the required dimensions for the design layout specifications. In addition, even though the Confidence Power Plus Treadmill weight capacity is only 250 lbs, the treadmill deck will either be either replaced or reinforced during the assembly of the TwAS.

The design criteria for TwAS will be height of the treadmill, belt dimensions, treadmill speed, and the speed of stiffness adjustment. The height should be designed in such a way that the lower extremities of the person walking on the treadmill will remain within the space envelope that is trackable by our Vicon motion capture system. Considering the ceiling height of our laboratory (around 3 m) and average height of the subjects, the maximum height of the treadmill should be less than 1.2 m. Speed of the treadmill should be adjusted based on the speed of the person walking on the treadmill. The length of the belt should be large enough to account for small differences between walking speed of the person and the belt's speed. Therefore, in case there is any difference between the two speeds, the person would still remain on the belt before the belt's speed is adjusted. The width of the belt should give enough space for the person to step on the treadmill with both feet. The distance between left and right belts should be minimal to avoid any undesired effect on the walking gait in the frontal plane. Range of stiffness in our system should span from extremely rigid, which can represent a normal rigid surface, to any level, even completely passive. The completely passive case would be useful to study fall reactions in people when suddenly the surface stiffness drops to zero.

The TwAS is composed of two identical parts: the associated left leg and the associated right leg. Each part has a treadmill whose speed can be controlled independently using a dedicated motor and is composed of two modules: the force transmission module and the stiffness adjustment module. When a person steps on the treadmill, the compliant surface experiences a vertical displacement due to the applied vertical force. The force transmission module transfers this vertical displacement of the treadmill to a horizontal movement of an input link as shown in Fig. 7.22. In order to constrain the displacement of the treadmill to a purely vertical motion without impeding the human motion, we use a scissor lift mechanism (see Fig. 7.22).

The scissor lift has two scissor arms intersecting at a pinned joint that allows them to rotate, relatively. The first arm is connected to the treadmill surface from one end through a fixed revolute joint and can slide on the holding frame at the other point through a roller bearing. This end of

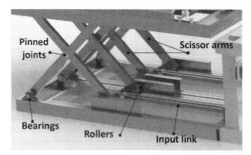

Figure 7.22 Force transmission module of TwAS: vertical displacement of and force applied at the treadmill surface will be transferred to horizontal displacement and force of the input link through rotation of the diagonal scissor arms around a pinned joint.

the first arm is connected to the input link. The second arm, in contrast, is connected to the holding frame from one end through a fixed revolute joint and can slide under the treadmill surface at the other point through a roller bearing. As the treadmill surface moves downward due to the vertical force, the two arms rotate around their intersecting pinned joint and their two ends move away from each other. As a result, the input link moves out horizontally toward the front side and transfers the vertical force to a horizontal force, as shown in Fig. 7.23.

The input link can move through a guide that is composed of linear bearings and constrains its motion to a horizontal movement. From the other end, the input link is connected to the stiffness adjustment module. Therefore it can transmit the resultant horizontal force to the stiffness adjustment module.

The stiffness adjustment module, as shown in Fig. 7.23, is composed of a lever, a pivot point, and four springs. One end of the lever is connected to the input link through a sliding joint and from the other end to the spring holder through a roller joint. The lever can rotate around a pivot point, which is a roller bearing mounted on the nut of a ball-screw mechanism. The motion of the spring holder is also constraint through a horizontal pipe mounted inside the springs and welded to the holding frame of the stiffness adjustment module. This imposed constraint allows for decoupling the resultant surface stiffness from the surface displacement, as the force due to the spring deflection is always guaranteed to act upon the lever completely horizontally.

As the input link pushes the lever to the right side, the lever rotates around the pivot point in a counterclockwise direction. The rotation of the lever around the pivot point will force the other end of the lever to the left side. As a result, the spring holder will move horizontally to the left side and compress the springs. Springs are placed in two parallel rows,

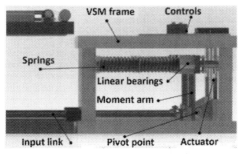

Figure 7.23 Stiffness adjustment module of TwAS: input link transfers the force to the lever, which rotates around a pivot point and deflect the spring.

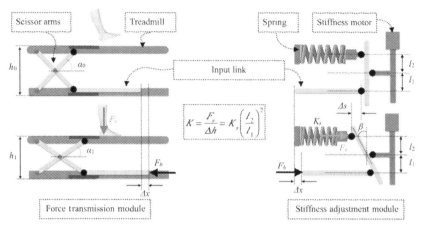

Figure 7.24 Schematic of stiffness adjustment of TwAS.

where each row is composed of two springs placed in series. This arrangement of the springs will result in a stiffness value equal to the stiffness of a single spring, whereas the overall allowable deflation will be doubled. The pivot point of the lever is connected to a ball screw mechanism that is powered by a dedicated stiffness adjustment actuator. Therefore, the pivot point can travel along the side of the lever from the spring holder point (i.e., spring point) to the input link attachment (i.e., force point). The vertical movement of the pivot point between the spring and force points will alter the effective stiffness of the lever (Fig. 7.24).

As the structure of the treadmill will be subjected to loading and unloading repeatedly, a force analysis is required to make sure that the system is stable and robust to fatigue. Through extensive and detailed finite element analysis (FEM) analysis, two components were noticed to be critical: one is the scissor arm that will experience the highest deflection and the other one is the pivot holder (that will be connected to the ball screws nut) that will receive the highest pressure.

7.3 Experimental results for stiffness adjustment of Treadmill with Adjustable Stiffness

As mentioned before, the advantages of TwAS compared to other similar systems are as follows: (1) purely vertical displacement of the surface; (2) wide range of stiffness regulation; (3) the ability to bilaterally regulate the surface stiffness; (4) surface stiffness consistency, independent of the location of the person on the treadmill.

In addition, the surface stiffness with the TwAS system is independent of the surface displacement. The experiments presented in this section are designed to prove these specifications. Furthermore, some preliminary experiments are presented to show the effects of surface stiffness on the energy expenditure of walking and joints trajectories.

Before we discuss each experiment, we briefly explain different components of our system, beside the TwAS treadmill. In order to guarantee the safety of the person walking over the treadmill, the LiteGait harness system has been integrated with the TwAS. This harness system can also provide weight-bearing capacity to partially cancel the weight of the person up to 130 kg [14].

To measure the force, ATMI force plates are used that can measure up to 4450 N vertical force [8]. With the use of Vicon motion capture systems with eight Xeon cameras (@3.7 GHz), the gait of the person is tracked [26]. The Vicon system is commercially integrated with the ATMI force plates.

Finally, the energy expenditure of the person is measured by using the ParvoMedics VO_2 system [27]. Fig. 7.25 shows an overall picture of TwAS with its integrated systems.

7.3.1 Range of stiffness adjustment

As mentioned before, theoretically the TwAS treadmill can change the stiffness from very soft to very rigid. The maximum stiffness that can be expected is therefore the structural stiffness of the system. In order to

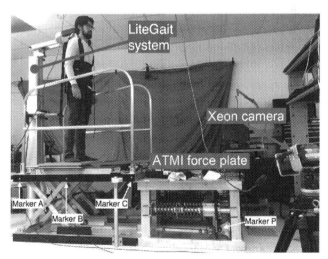

Figure 7.25 Apparatus to measure surface stiffness of the TwAS treadmill.

show the range of actual surface stiffness that can be realized with TwAS treadmill, we performed the following experiments. The ATMI force plate was placed over the treadmill surface. Then, a subject with weight around 75 kg stood on the treadmill while he was strapped to the harness system.

Three markers were placed along the treadmill surface (markers A, B, and C). In addition, another marker (C) is placed at the pivot point. The markers A, B, and C form the skeleton of the treadmill surface in the sagittal plane. To measure the surface displacement, these markers were being tracked by the motion capture system.

The slope of each curve represents the stiffness of the surface. Dotted lines show expected forces. In order to measure the stiffness and its range, first the pivot point was moved to the force point (the maximum stiffness). With the use of hydraulic system of the harness system, the weight of the person was being canceled from 0% to 100%. At each cancellation rate, the vertical force applied to the force plate (=actual weight − canceled weight) was being measured by the force plate and the surface displacement was being tracked by the motion capture system. Then, using the stiffness motor, the pivot was moved to different points and the abovementioned experiment was repeated. In total, the stiffness measurement experiments were conducted for six different points (i.e., level of surface stiffness) by changing the position of the pivot joint from the force point all the way up to the next end, close to the spring point. The results are shown in Fig. 7.26.

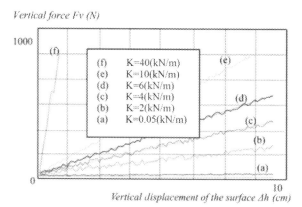

Figure 7.26 Changes in the surface displacement Δh as the result of canceling the weight F_v of the subject with LiteGait system. The slope of each curve represents the stiffness of the surface. Dotted lines show expected forces.

By connecting the vertical forces corresponding to surface displacements for each level of stiffness, Fig. 7.26 reveals two important features of the TwAS treadmill: first, a wide range of surface stiffness (slope of each line) can be achieved; and second, stiffness is decoupled from the displacement (each level of stiffness represents a line), so by setting the stiffness to a certain value, it is guaranteed that it will remain constant, independent of the external force.

In addition, expected stiffnesses for each position of the pivot point based on stiffness adjustment mechanism, are also plotted in Fig. 7.26 (dotted lines), which shows the accuracy of our model in predicting the surface stiffness

7.3.2 Independence of the surface stiffness with respect to location of the person and vertical displacement of the surface

In this experiment, the surface stiffness is first set to a certain compliant level. Then, a person stands statically at three different locations on the surface: two extreme locations at each end, and one at the middle of the treadmill, as shown in Fig. 7.27.

During the loading process at each location, the trajectories of markers A, B, and C in sagittal plane were being tracked by the motion capture system, as shown in Fig. 7.28.

As Fig. 7.27 shows, the surface displacement remained unchanged for the three locations, which reveals that the surface stiffness of the TwAS treadmill is independent of the location of the person on the treadmill. Again, this is a very helpful feature when the speed of the treadmill is not exactly matched with the speed of locomotion, which would result in a relative motion between the person and the treadmill.

Figs. 7.27 and 7.28 also reveal another important feature of the TwAS system when considering the trajectories of markers A, B, and C, which is no matter what the location of the person, the surface displacement is

Figure 7.27 Consistency of the surface stiffness regardless of the location of the person.

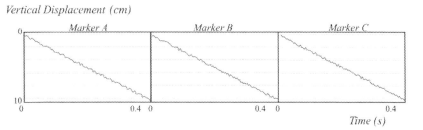

Figure 7.28 Trajectories of markers A, B, and C during displacement of the surface.

purely vertical. By calculating the position of each marker at each instance of the time, it was found that the surface inclination angle remained within the range of $+0.003$ to -0.002 degrees. Therefore the displacement of the surface does not impose any unwanted kinematic constraint on the motion of the ankle joint as is the case with the variable stiffness treadmil system [24]. This is because the scissor mechanism of the TwAS system restricts the surface displacement to a purely vertical one.

7.3.3 Bilateral stiffness adjustment

As mentioned before, the TwAS system consists of two independent but identical left and right treadmills. To prove the bilateral stiffness adjustability, a person stepped on the system with his right leg is on the right treadmill and left leg on the left treadmill. The person was asked to start walking at normal speed where the speeds of both treadmills were adjusted accordingly. Initially both treadmills were set to maximum surface stiffness. Then, the surface stiffness of the right leg gradually decreased to its minimum level while the other treadmill was still set to maximum stiffness. As a result, the right treadmill experienced an increased surface displacement, which led the person to adjust his gait accordingly. Then, the surface stiffness of the right treadmill started to increase gradually up to its maximum. Once it reached the maximum stiffness level, the left treadmill started to gradually decrease its surface stiffness to its minimum, which again led to an asymmetric gait for the person. Then, again the surface stiffness of the left treadmill started to gradually increase to its maximum. Once it reached its maximum, then both treadmills started to gradually and equally decrease their surface stiffness to the minimum level. This again led the person to adjust his gait accordingly though the gait remained symmetrical.

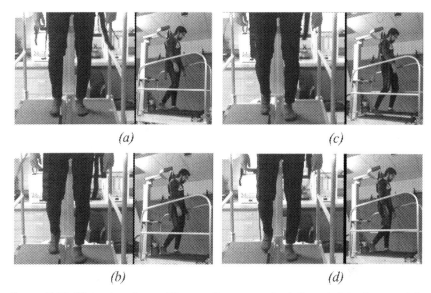

Figure 7.29 Bilateral surface stiffness adjustments for left and right legs, and its effect on the walking gait and posture. Snapshots are taken at the right leg toe-off moments. (A) When both surfaces are rigid; (B) when both surfaces are compliant; (C) when right surface is compliant and left is rigid; and (D) when the left surface is compliant and right is rigid.

Fig. 7.29 shows the front- and side-view snapshots of the person performing walking at the toe-off moment for right leg in the following four different cases:
1. surface stiffness of both treadmills at a low compliant level (see Fig. 7.29A);
2. surface stiffness of both treadmills at a high compliant level (see Fig. 7.29B);
3. surface stiffness of the right treadmill at a low level while that of the left treadmill remains at maximum level (see Fig. 7.29C); and
4. surface stiffness of the left treadmill at a low level while that of the right treadmill remains at maximum level (see Fig. 7.29D).

As Fig. 7.29 shows, the TwAS system is completely capable of adjusting the surface stiffness bilaterally for left and right legs. This is an extremely helpful feature to help people who have asymmetric gaits due to a disease or trauma.

7.3.4 Effects of surface stiffness on human gait and metabolic cost

We conducted some preliminary experiments in order to show that the surface compliance can affect the human gait (i.e., joint trajectories) and the energy consumption. The joint trajectories were being tracked using our motion capture system while the markers were placed on the subject as well as on the treadmill surface to track its displacement during the gait cycles. In order to measure the energy consumption of the human, we used our VO_2max system that can measure the volume of input oxygen and output carbon dioxide as an indicator for the metabolic cost. The VO_2max system was synchronized with the treadmill and motion capture system. We then asked a healthy subject to walk normally over the treadmill with the highest stiffness ($K = 40{,}000$ N/m). We then captured the joint trajectories with motion capture system and measured the metabolic cost with the VO_2max system. We then adjusted the surface stiffness to $K = 5000$ N/m and asked the person to walk over the treadmill with the same speed. The joint trajectories and metabolic cost were again tracked and measured using motion capture and VO_2max systems, respectively. Finally, we asked the person to walk at a reduced speed (as was preferred by the user) on the treadmill with stiffness $K = 5000$ N/m while joint trajectories and metabolic cost were tracked and measured. Fig. 7.30 shows the knee flexion trajectories of the subject in performing these three experiments.

The knee flexion trajectories were averages over 20 complete gate cycles. As is clear from Fig. 7.30, surface stiffness can greatly affect the knee flexion trajectory, especially during the stance phase. However, when walking with decreased speed over the compliant surface, the ankle flexion trajectory tended to its normal walking trajectory over a rigid surface. The surface compliance affected the knee angle at the heel strike, as

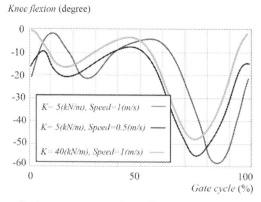

Figure 7.30 Knee flexion trajectories for different surface stiffness and walking speed.

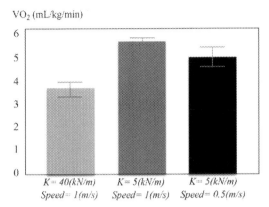

Figure 7.31 Oxygen consumption for different surface stiffness and walking speed.

the corresponding leg touched the surface while the knee was still bent. This is due to the fact that the treadmill surface associated with the other leg is deflected when the corresponding leg touches the treadmill surface.

We also measured the oxygen consumption of the subject in performing these experiments using the VO_2max system. Fig. 7.31 shows the result for normal walking speed over a rigid surface, normal walking speed over a soft surface, and slow walking over a soft surface.

As Fig. 7.31 shows, the minimum metabolic cost belongs to normal walking speed over a rigid surface. As the surface compliance increased, the subject required more oxygen consumption to perform the task. However, when the subject decreased the walking speed to half the normal speed the energy consumption reduced. The results clearly show that surface compliance can affect the energy expenditure of the locomotion. Fig. 7.14 shows the experimental setup.

7.3.5 Kinematics of human walking on Treadmill with Adjustable Stiffness

Encountering walking surfaces with different stiffness properties, such as sand, grass, mud, carpets, or even foams, is unavoidable in our daily life [26–38].

Sometimes, people are even prescribed to walk on compliant surfaces, such as mobility impaired patients (Hirsch et al., 2003; Cho and Lee, 2010) or injured athletes (Smith, 2006; Impellizzeri et al., 2008). Walking on compliant surfaces affects the gait kinematics (Menant et al., 2009), dynamics, muscle activation patterns (Harput et al., 2013) and metabolic cost of walking (Zamparo et al., 1992). (MacLellan and Patla, 2006a) discussed the adaptation of walking pattern on a compliant surface to regulate the dynamic stability. It was found that the vertical motion of center of mass COM decreased on the compliant surface to provide a more stable posture when walking on the surface while the mediolateral COM was not affected. Toe trajectory during the swing phase was also elevated to avoid tripping on the deformable compliant surface. Step width and length increased on the compliant surface to increase base of support and provide better control of COM.

In another study (MacLellan and Patla, 2006b), the role of the central nervous system (CNS) in adopting different strategies to step over obstacles during walking on a compliant surface was presented and compared with the same scenario on rigid surfaces. In this study, an increase in gastrocnemius and soleus activity during push-off accounted for increases in step length seen on the compliant surface. Dynamic stability margin in the anterior–posterior direction demonstrated a constant overcompensation and subsequent correction in COM control.

Chang et al. [31] compared the ability of some measurement tools, such as stride interval dynamics (Hausdorff et al., 1997) and the maximum Lyapunov exponent (Bruijn et al., 2012), to detect changes between over ground and compliant surface walking, a condition known to affect stability, to determine their aptness as measures of dynamic stability. Barbara et al. [28] compared kinematic gait data between two groups of young and old adults to perform the Timed Up and Go Test (TUG) (Shumway-Cook et al., 2000) while walking on a rigid and on a compliant surface. They found that the elder group required more time to complete the task, however, the gait speed, stride length, and vertical displacement of the foot were similar for both groups. The compliant surface also reduced the walking speed in both groups.

Another study [27] compared walking speed, step width, step time asymmetry, step time variability, and the speed and accuracy of cognitive task performance in performing a dual-task while walking on a compliant surface between transfemoral amputees (TFA) and a control group. No significant group-by-task interactions were observed for cognitive task performance. A significant group-by-task interaction for step time asymmetry indicated that participants with TFA increased temporal asymmetry in dual-task relative to single-task conditions, while control participants maintained symmetrical gait.

In the abovementioned researches as well as all other similar researches, understanding the effect of a compliant surface on the walking performance has been limited to one compliant surface and comparing the gait data with walking on a rigid surface. However, so far, no study has sought to understand how the level of surface stiffness plays a role in altering the abovementioned gait parameters and to what extent these gait parameters change as a function of surface stiffness. Such understanding will have a great impact on the rehabilitation of mobility impaired patients or the recovery of injured athletes. For example, determining what level of surface stiffness is best for an individual poststroke patient to maximize the benefit of the rehabilitation process or to speed up recovery of an injured athlete.

In this work, and as the first step toward understanding the role of surface stiffness on the gait performance, the effect of level of surface stiffness on the trajectories of hip, knee, and ankle during walking at a normal walking speed 1 m/s is studied.

7.3.5.1 Method
Participants: 18 healthy adults (11 males and 7 females, 20.3 ± 5.8 years old, 73.2 ± 10.2 kg, and 178.0 ± 11.3 cm). The experimental protocol has been approved by the Institutional Review Board (IRB) office at the University of Texas at San Antonio and all participants received informed consent on the experimental protocol prior to conducting the experiments.

7.3.5.2 Procedures and instrumentation
In order to accurately adjust the surface stiffness level, the TwAS treadmill has been integrated with a LiteGait (LiteGait: The Original Partial-Weight-Bearing Gait Therapy Device, n.d.) harness system that will hold the user if (s)he is going to fall to prevent any injury.

Each participant had a session that included walking at a speed of 1 m/s on surfaces with six different stiffness levels as follows: $K_1 = 10$ MN/m, $K_2 = 1$ MN/m, $K_3 = 100$ kN/m, $K_4 = 10$ kN/m, $K_5 = 5$ kN/m, and $K_6 = 3300$ N/m. Before each trial, the level of surface stiffness was calibrated by attaching markers on the treadmill's belt and tracking their vertical downward motion due to a certain weight placed on top of the belt.

On each compliant surface, subjects walked for at least 5 min, where joint kinematics of the last 2.5 min were measured using a Vicon motion capture system comprising eight cameras (VICON Motion Capture System, n.d.) with a capture rate of 100 Hz. Following the procedure used by Voloshina et al. (2013), for all walking trials we recorded the position of reflective markers located on the pelvis and lower limbs. Markers were taped to the skin or spandex-like shorts worn by the subjects. Vicon Track Manager and 3D motion analysis software (Nexus) were used to process, filter, and calculate lower extremity joints' trajectories from the position data. Furthermore, for each participant, leg length was calculated as the distance between the marker attached to the pelvis and the one attached to the ankle joint. Participants walked on the treadmill without wearing shoes in order to eliminate the effect of the stiffness of the shoes on the kinematics data.

Statistics: Stride cycle was defined as the interval between two consecutive heel strike moments of the same leg. For each subject, joint trajectories we averaged for each level of surface stiffness. Means were calculated across trials for each point in relative stride cycle timing. Similarly, joint trajectory variability was defined for each subject and level of surface stiffness as the standard deviation across trials for each point. We then reported the mean variations (and standard deviations) across subjects for each surface stiffness value. The trajectories trends were evaluated qualitatively.

7.3.5.3 Results

Figs. 7.32−7.34, shows the joint trajectories during one full walking cycle for different levels of surface stiffness (K_1 to K_6). The vertical axes represent joint angles where positive values are the joints' extension angles and negative values are the joints' flexion angles in the sagittal plane. As shown in Fig. 7.32, although the trajectory of the hip joint changes in magnitude as the level of surface stiffness is adjusted to different levels, its trend remains almost the same: the hip angle starts from a negative value (flexed), then it increases to a maximum value (still flexed), and again decreases to further flexion. However, as the surface becomes more compliant, increases in the standard deviation of the hip trajectory among the subjects can be clearly seen.

Starting from a rigid surface K_1, the trajectory of the knee joint (Fig. 7.33) changes in magnitude but holds almost the same trend down to a certain level of surface stiffness K_5. At high levels of surface stiffness, the hip trajectory starts at a flexed configuration while its angle slightly decreases to a more flexed configuration and then increases to maximum value (always equal to or less than zero) at the toe-off moment. During the flight phase, the knee angle experiences a large flexion from this maximum point to a minimum point (maximum flexion) and then goes back

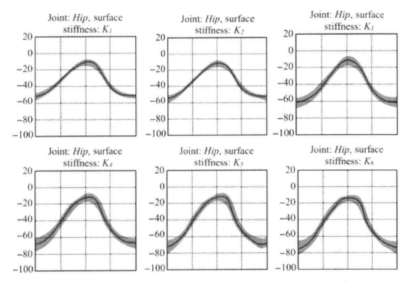

Figure 7.32 Trajectories of the hip joint while walking at the speed of 1 m/s on surfaces with six different levels of stiffness.

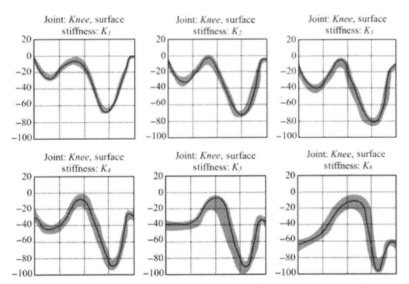

Figure 7.33 Trajectories of the Knee joint while walking at the speed of 1 m/s on surfaces with six different levels of stiffness.

to its initial flexed position. However, on very compliant surfaces, the knee joint starts from a way more flexed configuration and then monotonically increases to a maximum value before the toe-off moment.

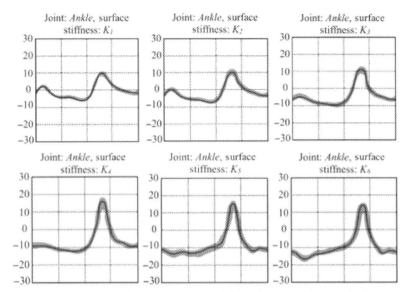

Figure 7.34 Trajectories of the ankle joint while walking at the speed of 1 m/s on surfaces with six different levels of stiffness.

In addition, as the surface becomes more compliant, the standard deviation of the knee trajectory considerably increases among the subjects.

Similarly, the trajectory of the ankle joint (Fig. 7.34) changes in magnitude as the surface becomes more compliant while its trend remains almost the same down to the K_5 level of surface stiffness. The change in the ankle's trajectory is more obvious during the stance phase of walking, especially at its beginning. Right after the heel strike moment, on a rigid surface the angle of the ankle joint slightly increases and then decreases to a minimum at the toe-off moment. However, on a very compliant surfaces, the ankle's angle first decreases to a minimum value soon after the heel-strike moment and then monotonically increases during the stance phase. The standard deviation of the ankle trajectory, however does not significantly change among the participants as the level surface stiffness decreases.

7.3.5.4 Discussion

In this study, the effect of altering surface stiffness on the trajectories of hip, knee, and ankle joints in the sagittal plane was examined for 18 healthy adults. In all cases, changing the surface stiffness affected these trajectories in magnitude but their trend remained almost the same down to a threshold surface stiffness value of 5 kN/m. Walking on surfaces softer than this level changed the trajectory trends, especially for the ankle and the knee joints.

Once the surface is very compliant, its deflection (downward displacement) due to the normal force will be considerably large. Therefore during the stance phase of one leg and the flight phase of the other leg, one belt experiences a large deflection while the other belt is not deflected, resulting in a large vertical distance between the surfaces of the two belts. Therefore our hypothesis is that at the heel strike moment of the other leg (the one that has been in the flight phase) the legs' configuration pretty much tends to be the one that humans take during stair climbing.

Comparing the knee's and hip's trajectories on very compliant surface K_6, in Figs. 7.32–7.34, with those presented by Afzal et al. (2013) and Riener et al. (2002), one can clearly see the similarities between these trajectories while walking on very compliant surface and stair climbing. During stairs negotiation, the step height as well as the human height play important roles in affecting the joint trajectories. Similarly, we believe what, in fact, affected the joint trajectories in our experiments is not the surface stiffness parameter alone, but the amount of surface deflection due to its compliance and its relationship to the length of the subjects' legs. We measured the leg length of each participant as the distance between the pelvis and the ankle joint as well as maximum surface deflection while the participant was walking on the very compliant surface, using the motion capture systems.

We then compared the two variables (i.e., the leg length and maximum surface deflection) for each participant. The results are plotted in Fig. 7.35. Performing a correlation test on these data resulted in a correlation coefficient of 0.73, which suggests there is a high dependency ratio between the height of the human, the weight (which affects the surface deflection), and the threshold level for the surface stiffness at which the trends of joint trajectories would change.

In addition, as the surface became more compliant, we clearly noticed an increase in joint trajectories' standard deviations. That could be due to different strategies that can be taken by participants while walking on compliant surfaces, some may manifest themselves in a frontal plane rather than sagittal plane. However, further studies are required to confirm this hypothesis.

Finally, even though walking on very compliant surfaces may tend to be similar to stair negotiation in terms of joint trajectories, we believe that

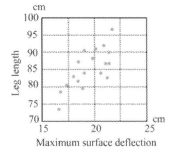

Figure 7.35 Leg length versus maximum deflection of the surface while walking on compliant surface with threshold stiffness level $K_5 = 5$ kN/m. The correlation coefficient is 0.73.

would not be the case regarding other gait parameters, such as joint dynamics, muscle activation patterns, and metabolic cost. In future work, we will study the effect of surface stiffness on these parameters, as well as joint trajectories in the frontal plane, especially the trajectory of the hip joint.

References

[1] M.G. Catalano, G. Grioli, M. Garabini, F. Bonomo, M. Mancini, N. Tsagarakis, et al., VSA-CubeBot: a modular variable stiffness platform for multiple degrees of freedom robots, 2011 IEEE International Conference on Robotics and Automation, IEEE, 2011, pp. 5090–5095.

[2] A. De Luca, F. Flacco, A. Bicchi, R. Schiavi, Nonlinear decoupled motion-stiffness control and collision detection/reaction for the VSA-II variable stiffness device, 2009 IEEE/RSJ International Conference on Intelligent Robots and Systems, IEEE, 2009, pp. 5487–5494.

[3] M. Azadi, S. Behzadipour, G. Faulkner, Antagonistic variable stiffness elements, Mech. Mach. Theory 44 (2009) 1746–1758.

[4] C. English, D. Russell, Implementation of variable joint stiffness through antagonistic actuation using rolamite springs, Mech. Mach. Theory 34 (1999) 27–40.

[5] F. Petit, M. Chalon, W. Friedl, M. Grebenstein, A. Albu-Schäffer, G. Hirzinger, Bidirectional antagonistic variable stiffness actuation: analysis, design & implementation, 2010 IEEE International Conference on Robotics and Automation, IEEE, 2010, pp. 4189–4196.

[6] J.W. Hurst, J.E. Chestnutt, A.A. Rizzi, An actuator with physically variable stiffness for highly dynamic legged locomotion, IEEE International Conference on Robotics and Automation, 2004. Proceedings. ICRA'04. 2004, IEEE, 2004, pp. 4662–4667.

[7] William Townsend, Kenneth Salisbury, Mechanical bandwidth as a guideline to high-performance manipulator design, Proceedings, 1989 International Conference on Robotics and Automation, Scottsdale, AZ, 3 (1989) 1390–1395. Available from: https://doi.org/10.1109/ROBOT.1989.100173.

[8] S. Kianzad, M. Pandit, J.D. Lewis, A.R. Berlingeri, K.J. Haebler, J.D. Madden, Variable stiffness and recruitment using nylon actuators arranged in a pennate

configuration, Electroactive Polymer Actuators and Devices (EAPAD) 2015, International Society for Optics and Photonics, 2015, p. 94301Z.
[9] M. Garabini, A. Passaglia, F. Belo, P. Salaris, A. Bicchi, Optimality principles in variable stiffness control: the VSA hammer, 2011 IEEE/RSJ International Conference on Intelligent Robots and Systems, IEEE, 2011, pp. 3770–3775.
[10] R. Schiavi, G. Grioli, S. Sen, A. Bicchi, VSA-II: a novel prototype of variable stiffness actuator for safe and performing robots interacting with humans, 2008 IEEE International Conference on Robotics and Automation, IEEE, 2008, pp. 2171–2176.
[11] G. Tonietti, R. Schiavi, A. Bicchi, Design and control of a variable stiffness actuator for safe and fast physical human/robot interaction, Proceedings of the 2005 IEEE International Conference on Robotics and Automation, IEEE, 2005, pp. 526–531.
[12] A. Albu-Schäffer, S. Wolf, O. Eiberger, S. Haddadin, F. Petit, M. Chalon, Dynamic modelling and control of variable stiffness actuators, 2010 IEEE International Conference on Robotics and Automation, IEEE, 2010, pp. 2155–2162.
[13] T.-H. Huang, J.-Y. Kuan, H.-P. Huang, Design of a new variable stiffness actuator and application for assistive exercise control, 2011 IEEE/RSJ International Conference on Intelligent Robots and Systems, IEEE, 2011, pp. 372–377.
[14] A. Jiang, G. Xynogalas, P. Dasgupta, K. Althoefer, T. Nanayakkara, Design of a variable stiffness flexible manipulator with composite granular jamming and membrane coupling, 2012 IEEE/RSJ International Conference on Intelligent Robots and Systems, IEEE, 2012, pp. 2922–2927.
[15] C. Santulli, S.I. Patel, G. Jeronimidis, F.J. Davis, G.R. Mitchell, Development of smart variable stiffness actuators using polymer hydrogels, Smart Mater. Struct. 14 (2005) 434.
[16] I. Sardellitti, G.A. Medrano-Cerda, N. Tsagarakis, A. Jafari, D.G. Caldwell, Gain scheduling control for a class of variable stiffness actuators based on lever mechanisms, IEEE Trans. Robot. 29 (2013) 791–798.
[17] T.W. Secord, H.H. Asada, A variable stiffness PZT actuator having tunable resonant frequencies, IEEE Trans. Robot. 26 (2010) 993–1005.
[18] L.C. Visser, R. Carloni, R. Ünal, S. Stramigioli, Modeling and design of energy efficient variable stiffness actuators, 2010 IEEE International Conference on Robotics and Automation, IEEE, 2010, pp. 3273–3278.
[19] M.C. Yuen, R.A. Bilodeau, R.K. Kramer, Active variable stiffness fibers for multifunctional robotic fabrics, IEEE Robot. Autom. Lett. 1 (2016) 708–715.
[20] A. Jafari, H.Q. Vu, F. Iida, Determinants for stiffness adjustment mechanisms, J. Intell. Robot. Syst. 82 (2016) 435–454. Available from: https://doi.org/10.1007/s10846-015-0253-8.
[21] A. Jafari, Coupling between the output force and stiffness in different variable stiffness actuators, Actuators 3 (2014) 270–284. Available from: https://doi.org/10.3390/act3030270.
[22] A. Jafari, N.G. Tsagarakis, I. Sardellitti, D.G. Caldwell, How design can affect the energy required to regulate the stiffness in variable stiffness actuators, 2012 IEEE International Conference on Robotics and Automation. Presented at the 2012 IEEE International Conference on Robotics and Automation, IEEE, 2012, pp. 2792–2797. Available from: https://doi.org/10.1109/ICRA.2012.6224946.
[23] A. Jafari, N.G. Tsagarakis, D.G. Caldwell, AwAS-II: a new actuator with adjustable stiffness based on the novel principle of adaptable pivot point and variable lever ratio, 2011 IEEE International Conference on Robotics and Automation. Presented at the 2011 IEEE International Conference on Robotics and Automation, IEEE, 2011, pp. 4638–4643. Available from: https://doi.org/10.1109/ICRA.2011.5979994.

[24] A. Jafari, N.G. Tsagarakis, D.G. Caldwell, Exploiting natural dynamics for energy minimization using an Actuator with Adjustable Stiffness (AwAS), 2011 IEEE International Conference on Robotics and Automation. Presented at the 2011 IEEE International Conference on Robotics and Automation, IEEE, 2011, pp. 4632−4637. Available from: https://doi.org/10.1109/ICRA.2011.5979979.
[25] A. Jafari, N.G. Tsagarakis, D.G. Caldwell, AwAS-II: a new actuator with adjustable stiffness based on the novel principle of adaptable pivot point and variable lever ratio, 2011 IEEE International Conference on Robotics and Automation, IEEE, 2011, pp. 4638−4643.
[26] O. Mohamed, K. Cerny, W. Jones, J.M. Burnfield, The effect of terrain on foot pressures during walking, Foot Ankle Int. 26 (2005) 859−869.
[27] S.J. Morgan, B.J. Hafner, V.E. Kelly, Dual-task walking over a compliant foam surface: a comparison of people with transfemoral amputation and controls, Gait Posture 58 (2017) 41−45.
[28] R. Bárbara, S.M. Freitas, L.B. Bagesteiro, M.R. Perracini, S.R. Alouche, Gait characteristics of younger-old and older-old adults walking overground and on a compliant surface, Braz. J. Phys. Ther. 16 (2012) 375−380.
[29] J. Bjerke, F. Öhberg, K.G. Nilsson, A.-K. Stensdotter, Walking on a compliant surface does not enhance kinematic gait asymmetries after unilateral total knee arthroplasty, Knee Surg. Sports Traumatol. Arthrosc. 24 (2016) 2606−2613.
[30] C. Canudas, L. Roussel, A. Goswami, Periodic stabilization of a 1-DOF hopping robot on nonlinear compliant surface, IFAC Proc. 30 (1997) 385−390.
[31] M.D. Chang, E. Sejdić, V. Wright, T. Chau, Measures of dynamic stability: detecting differences between walking overground and on a compliant surface, Hum. Mov. Sci. 29 (2010) 977−986.
[32] M.H. Cole, P.A. Silburn, J.M. Wood, G.K. Kerr, Falls in Parkinson's disease: evidence for altered stepping strategies on compliant surfaces, Parkinsonism Relat. Disord. 17 (2011) 610−616.
[33] S.E.H. Davies, S.N. Mackinnon, The energetics of walking on sand and grass at various speeds, Ergonomics 49 (2006) 651−660.
[34] D.H. Gates, S.J. Scott, J.M. Wilken, J.B. Dingwell, Frontal plane dynamic margins of stability in individuals with and without transtibial amputation walking on a loose rock surface, Gait Posture 38 (2013) 570−575.
[35] F. Plestan, J.W. Grizzle, E.R. Westervelt, G. Abba, Stable walking of a 7-DOF biped robot, IEEE Trans. Robot. Autom. 19 (2003) 653−668.
[36] E.L. Radin, R.B. Orr, J.L. Kelman, I.L. Paul, R.M. Rose, Effect of prolonged walking on concrete on the knees of sheep, J. Biomech. 15 (1982) 487−492.
[37] L. Turchet, S. Serafin, P. Cesari, Walking pace affected by interactive sounds simulating stepping on different terrains, ACM Trans. Appl. Percept. TAP. 10 (2013) 1−14.
[38] Y. Visell, B.L. Giordano, G. Millet, J.R. Cooperstock, Vibration influences haptic perception of surface compliance during walking, PLoS One 6 (3) (2011) e17697.

CHAPTER 8

An artificial skeletal muscle for use in pediatric rehabilitation robotics

Ahad Behboodi[1,2], James F. Alesi[1] and Samuel C.K. Lee[1,2,3]
[1]Biomechanics and Movement Science Program, University of Delaware, Newark, DE, United States
[2]Department of Physical Therapy, University of Delaware, Newark, DE, United States
[3]Shriners Hospitals for Children, Philadelphia, PA, United States

8.1 Introduction

Soft actuators, that is, actuators having Young's modulus in the range of soft biological material ($10^4 - 10^9$ Pa) such as muscle and tendon (Fig. 8.1), are highly desirable in assistive robots because they increase the safety of the device and conform more readily to various objects [1]. When natural feel combines with acoustically noiseless operation, the resultant actuator may substantially increase the acceptability of the assistive robots that include them. These features may be even more significant for acceptability in pediatric populations. Additionally, a soft actuator that contracts longitudinally can more closely follow any complex origin and insertion point of a target muscle, for example, a supinator muscle, to augment its complex motion while occupying less space. Such continuously deformable muscle-like actuators can be configured to actuate virtually unlimited types of degrees of freedom in contrast to their hard-bodied counterparts [1,2].

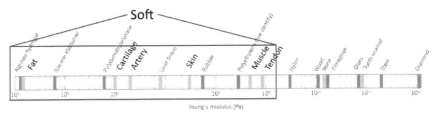

Figure 8.1 Softness in comparison to biological tissue [1]. The Young's modulus of biological tissues and nonbiological materials are depected in green and blue, respectively.

A variety of electronic, ionic, thermally, and photoactivated soft actuators with interesting muscle-like characteristics have been developed [3–5]. Due to their limited scalability, manufacturing, and structural complexity, however, only a few of them have been shown to be capable of generating comparable force to human skeletal muscle. In this chapter, we call this group of soft actuators artificial skeletal muscles.

Recently, polymer-based actuators have attracted great attention due to their unique mechanical characteristics, which can be tailor-made for targeted applications [6]. Many electroactive polymers (EAP) and thermally driven polymer actuators have shown promising capability to act as artificial skeletal muscle.

Thermally activated actuators usually have large strain and high force generation, which surpass human skeletal muscle [7–9]. The most studied and commercially available actuators of this kind are shape memory alloys (SMAs) [10], which due to their high Young's modulus (2×10^{10} Pa), are not considered soft (Fig. 8.1). Although usually in the soft range, shape memory polymers (SMPs) have limited force generation capability [11]. Conversely, thermally driven polymer fibers, for example, coiled nylon fiber (CNF) actuators [7], and phase-change actuators, such as the ethanol-based phase-change (EPC) [12], are soft and have linear operation. Their large strain and power-to-weight ratios make them potentially versatile artificial skeletal muscles. Although CNF and EPC have high power-to-mass ratios, as is typical for thermally driven actuators, their relatively long response times, activation frequencies, and efficiencies limit their suitability as artificial skeletal muscle.

The activation mechanism is a prominent criterion in choosing a suitable actuator for any application. For active exoskeletons, light [13] and magnetically [14] activated actuators are less desirable because their source of activation decreases their power-to-mass ratio and complicates their structural design. Electrical stimulation is considered the most promising activation mechanism, because of its high speed, widespread availability, and advanced control systems [15]. Thus even thermally driven actuators use electrothermal activation [7,12]. Furthermore, since natural muscles are activated by electrochemical signals, for biomimetic applications and artificial muscles electrical activation may be more preferable. Therefore electroactive polymer (EAP) actuators, which are an emerging class of electromechanical transducers, are gaining substantial attention.

EAP actuators possess a unique combination of functional and structural properties that are unrivaled among actuation technologies. These properties include large active strains, high power density, high mechanical compliance, scalability, acoustic noiselessness, and—in most cases—affordability [4]. EAPs are commonly classified into two major families: ionic EAPs and electronic EAPs. Ionic EAP actuation mechanisms are mostly based on ion commute between two electrodes. Therefore EAPs usually need to operate either in a gel or a wet-state electrolyte [3]. The performance of the latter, for example, conductive polymers such as carbon nanotube (CNT) and ionic polymeric metallic/composites (IPMC), is adversely affected by their low efficiency, short life span, and high response time [3,8]. Additionally, the need for a wet state electrolyte limits their operating temperature and increases their structural and packaging complexity. On the other hand, plasticized poly vinyl chloride (PVC) gel actuators are inexpensive ionic EAP actuators, with strain, efficiency, and power-to-mass ratio surpassing many artificial skeletal muscles alternatives [9,16].

As opposed to those activated via charge or mass transportions, electronic EAPs (also known as "dry" EAPs or field-activated EAPs) respond to applied electric fields by either reorientation of their internal structure (also known as electrostrictive polymers) or the electrostatic interaction of opposite charges, known as Maxwell pressure. Maxwell pressure is the electrostatic pressure between the oppositely charged electrodes of a capacitor. Although electronic EAPs require high activation electric fields (~ 150 MV/m), their prominent properties of large strains and stresses, fast response times, high strain-rates, and long lifetimes tend to outweigh this disadvantage [3,15]. Despite these outstanding features, few electronic EAPs are suitable to act as artificial skeletal muscle. For example, ferroelectric polymers are too mechanically rigid [9]. Liquid crystal elastomers, while capable of producing large strains ($\geq 100\%$), have slow response times and limited stress generation capability [4] The existing CNT aero gels have low force generation capability [4]. Alternatively, dielectric elastomer (DE) actuators are soft, acoustically noiseless, and lightweight with outstanding muscle-like features, which make them one of the most anticipated and studied soft actuators [2,3,17–19]. Power-to-mass ratios that can surpass that of human skeletal muscles, length self-sensing, and energy recuperation capability are among their unique features [2,18]. Their high driving voltages, however, have so far limited the availability of compatible electrical components that can be used with them in a design.

DE actuators consist of a thin elastomer film, which is coated on both sides with mechanically compliant electrodes forming a capacitor. When a high direct current (DC) voltage is applied to this compliant capacitor, the electrodes squeeze the elastomeric dielectric in the thickness direction. DE actuators have already shown some promise as artificial skeletal muscles for orthotic and prosthetic devices [19]. These actuators have been formed into cylindrical rolls that have strain, shape, and performance similar to natural skeletal muscles [20]. One such actuator was configured to be a similar shape to a biceps muscle and to act on a life-size skeletal arm [21]. Although showing great potential, only a few proposed DE actuator configurations have been shown to generate enough force to be deployed in an assistive upper extremity exoskeleton. Among those that have, the hydraulically amplified self-healing electrostatic (HASEL) actuators proposed by Acome's group [22] combine the power of hydraulic actuators and muscle-like behavior of DE actuators. HASEL actuators showed substantial amount of force generation and tensile actuation.

The stacked DE actuators proposed by Kovacs et al. [23] have a powerful and versatile DE actuator configuration that shows the most similarity to human skeletal muscle relative to other soft actuators. They feature multiple layers of thin enhanced DE actuators, each of which can be considered as a sarcomere within a myofibril. These capabilities led these actuators to become the first and only commercially available DE actuators that can be deployed as artificial skeletal muscle [24].

Despite many efforts to compare alternative actuators [3–6], to the best of our knowledge, this chapter is the first one to focus specifically on comparing soft and noiseless actuators with demonstrated force and linear displacement comparable to that of human skeletal muscle. In this chapter, the aforementioned thermally driven actuators (i.e., CNF and EPC actuators) and EAP actuators (i.e., PVC gel, HASEL, and stacked DE) are compared in their capabilities to be deployed as artificial skeletal muscles. Thereby, the most suitable actuator is identified for further investigation.

8.1.1 Actuation using Maxwell pressure

Because all the EAP actuators compared in this study, that is, PVC gel, HASEL, and stacked DE, used Maxwell pressure to actuate, it may be useful to explain this phenomena, briefly.

When an elastic dielectric is sandwiched between two mechanically compliant electrodes, by applying a DC voltage to the electrodes, the resultant Maxwell pressure, compresses the dielectric in the thickness direction and creates motion (Fig. 8.2). Eq. (8.1) shows the parabolic relation of the Maxwell pressure and voltage, where ϵ_r and ϵ_0 are the polymer and void permittivity, respectively, V is the applied voltage between electrodes, and d is the thickness.

8.2 Method
8.2.1 Inclusion criteria
The three major inclusion criteria in this study are (1) softness, defined as having a Young's modulus in the range of soft biological materials ($10^4 - 10^9$ Pa); (2) demonstrated ability to lift a tensile load equal or greater than the weight of a hypothetical 95th percentile 2-year old boy's forearm and hand (3.38 N); and (3) aco ustically noiseless operation. The weight of the forearm and hand was estimated by using WHO child growth standard [25] and Winter's anthropometric data [26].

8.2.2 Comparison measures
Each actuator was evaluated against human skeletal muscle using the measures described below. Each measure is indicated as high/moderate/low relative to human skeletal muscle (Table 8.1). Measures are high if they are greater than or equal to those of human skeletal muscle; moderate if they are lower than, but within an order of magnitude of, those of human skeletal muscle; and low if they are more than one order of magnitude smaller than that of human skeletal muscle (Table 8.1). An actuator is deemed comparable to skeletal muscle in a specific measure if they are

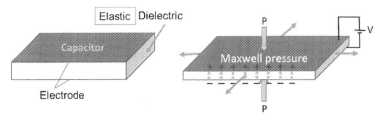

Figure 8.2 Maxwell pressure and the resultant compression or contraction of the elastic dielectric in a DEA compliant capacitor.

$$P = \varepsilon_r \varepsilon_0 \left(\frac{V}{d}\right)^2 \quad (8.1)$$

Table 8.1 High, moderate, and low range for the comparison measures.

Property	Skeletal muscle Typical	Skeletal muscle Max	Soft actuator (SA) High	Soft actuator (SA) Moderate	Soft actuator (SA) Low
Strain (%)	20	>40	≥ 20	$20 > SA > 2$	≤ 2
Stress (kPa)	100	350	≥ 100	$100 > SA > 10$	≤ 10
Strain-rate (%/s)		500	≥ 500	$500 > SA > 50$	≤ 50
Power-to-mass (W/kg)	50	200	≥ 50	$50 > SA > 5$	≤ 5
Efficiency (%)		40	≥ 40	$40 > SA > 4$	≤ 4

either moderate or high. Note that the typical human skeletal muscle values, reported in Table 8.1, first column, are used if available; otherwise, the maximum reported values (Table 8.1, second column) are used.

Unidirectional strain is defined as the ratio of unidirectional displacement to the initial length of the actuator.

Stress is the maximum generated force divided by cross-sectional area.

Strain-rate is a measure of actuation speed, and is defined as the maximum change in strain per unit of time. Unlike skeletal muscle, actuation frequency is reported as a measure of contraction speed for most of the actuators in this study. By assuming the actuation and recovery phase in a complete cycle have equal speed, we can derive a rough estimation of the strain-rate, for the sake of comparison. This assumption may not be correct for thermally driven actuators, where heating speed in the actuation phase may be higher than cooling speed in the recovery phase.

Power-to-mass ratio is maximum generated power normalized by mass.

Efficiency or electromechanical efficiency is the ratio of the mechanical (output) energy to the electrical (input) energy expended during a complete cycle of actuation.

8.3 Results

Five groups of actuators were chosen for the comparison: (1) CNF actuators; (2) EPC actuators; (3) PVC gel actuators; (4) stacked DE actuators; and (5) HASEL actuators. Although manufacturing cost and complexity were not part of the inclusion criteria, all of the compared actuators were moderately simple and inexpensive to make. Note that, not all the actuators in this study had reported value for power-to-mass ratio, strain-rate, and efficiency in the literature that we used.

8.3.1 Coiled nylon fiber actuators

These thermally driven actuators are made by coiling inexpensive high-strength, nylon threads. Heating an uncoiled nylon threads result in reversible thermal longitudinal contraction. For example, a straight nylon 6.6 thread with 127 μm diameter will contract 4% longitudinally when heated from 20°C to 240°C, which is similar to contractions achieved by commercially available NiTi shape memory wires [7]. Because longitudinal contraction is accompanied by radial expansion, these threads also function as torsional actuators. Thereby, the contraction can be amplified by twisting the thread to create coils (Table 8.2). The coiled actuator is made by twisting the thread while it is under tensile load and then thermally annealing to strengthen it. The amount of tensile load used in the coiling process dictates the spring index of the coil, that is, the ratio between the wire diameter and outer diameter of the wire coil. The larger the spring index, the smaller the strain and the larger the stress generation capabilities. Haines et al. [7] showed that after coiling, the thermal longitudinal contraction of nylon 6.6 increased from 4% (straight thread) to 34% (coiled thread) when heated from 20°C to 240°C. Additionally, by changing the direction of coiling, the resultant actuator fiber can either expand or contract by heating. When heated from 20°C to 120°C, a CNF with spring index of 1.7 (340 μm outer diameter) showed 21% unidirectional contraction and 22 MPa optimal load (for each fiber, the optimal load was defined as the load under which the maximum stress was realized), while a CNF with spring index of 1.1 (265 μm) showed 9.3% contraction and 50 MPa optimal load [7]. Interestingly, the CNF with spring index of 1.1 had an average power-to-mass ratio 42 times that of human skeletal muscle (2.1 kW/kg vs 50 W/kg, Table 8.1). Since this actuator has a very small diameter, it can be used in textiles; 12 CNF actuators were woven in parallel into a textile and could lift a 3 kg tensile load [8]. The cycle life of a CNF was 1.2 M cycles under 22 MPa of tensile load and actuation frequency of 1 Hz, and no significant change in unidirectional strain was observed [7]. When operating in air, the reported efficiency of these fibers was 1.08% [7]. Although 5 and 7.5 Hz actuation frequencies were reported in water and in helium actuation, respectively, the actuation frequency in open air was only 1 Hz (roughly 2%/s strain-rate) [7]. Table 8.2 summarizes the advantages and disadvantages of these actuators.

Table 8.2 The contraction of the coiled nylon 6.6 fiber is caused by an increase in temperature.

Advantages
High stress
High strain
Thin diameter
Disadvantages
Low strain-rate
Low efficiency
High driving temperature

By coiling the nylon thread, the longitudinal strain can be magnified substantially [7]. The right-hand column shows a summary of the advantages and disadvantages of these actuators.

Table 8.3 Structure and principle of operation of EPC actuators.

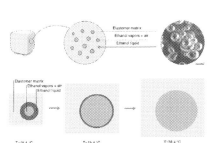

Advantages
High stress
High strain
3D printable
Disadvantages
Low strain-rate
Low efficiency
High power consumption

By increasing the temperature (T), the entrapped ethanol evaporates inside the elastomeric matrix and creates expansion [12]. Advantages and disadvantages of these actuators are summarized in the right-hand column.

8.3.2 Ethanol-based phase-change actuators

These phase-change actuators exploit the extreme volume change in liquid to vapor transition of ethanol and elastic properties of a silicone elastomer. By trapping ethanol in an elastomeric matrix, such as silicone, when heated to the boiling point of ethanol (78.4°C), the elastomer expands substantially due to the extreme volume change of ethanol's liquid to gas transition (Table 8.3). Force generation as high as 120 N (corresponding to 1.3 MPa of stress), for isometric contractions, and strain of 140%, under no load, were reported for a 2 g actuator [12]. The strain-rate and efficiency were extremely low, 2.5%/s and 0.2%, respectively [12]. A muscle-like configuration of EPC actuators was tested as a biceps muscle on a skeleton (without any control for the weight of the forearm) and could flex the arm to 90 degrees (the initial elbow angle was not reported). However, this motion required a substance amount of input power, 45 W. Also, the entrapped ethanol gradually escaped the elastomeric matrix through vaporization [12], which likely dramatically decreases the actuators' cycle life. The ethanol escape can be mitigated by special packaging to prevent evaporation. The advantages and disadvantages of this group of actuators can be seen in Table 8.3.

8.3.3 Poly vinyl chloride gel actuators

These gel-based ionic EAP actuators, known as PVC gel actuators, utilize Maxwell pressure to generate force. When sandwiched between two conductive surfaces, by applying electric field, they contract by Maxwell pressure. Due to their high elasticity, PVC gel actuators return to their initial

Table 8.4 Advantages and disadvantages of the actuation mechanism of a stacked PVC gel actuator.

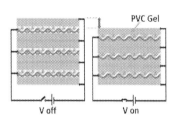

	Advantages
	Moderate stress
	Moderate strain
	Moderate strain-rate
	Power consumption
	Disadvantages
	Low power-to-mass ratio
	Pushing force
	Bulky and rigid packaging

When an electrical field is applied, PVC gel creeps into the air gap around the anode under Maxwell stress [9].

length when the electric field is removed (this is known as the recovery phase) (Table 8.4). Because the pushing force in the recovery phase, resulting from the gel's elasticity is significantly higher than the pulling force resulting from Maxwell pressure, the recovery phase is used as the actuation mechanism [9]. Under 400 V driving voltage, a rectangular actuator of 50 × 10 × 6 mm generated 39 N (78 kPa), 12% unidirectional contraction, 3 W/kg peak power-to-mass ratio, and 9 Hz bandwidth (roughly 240%/s) [9]. Power consumption under constant DC voltage of 400 V was 14.4 mW, which is low enough for potential wearable rehabilitation devices. Indeed, a PVC gel was tested as an artificial muscle to assist quadriceps muscle contraction on an exoskeleton [16]. However, cycle life testing showed a 64% reduction in contraction strain after about 170k cycles. Contraction performance remained constant from 170k cycles to the end of the testing (5.4 M cycles) [9]. Additionally, most of the proposed packaging for PVC gel actuators are rigid and bulky. The summary of the advantages and disadvantages of these actuators are presented in Table 8.4.

8.3.4 Stacked dielectric elastomer actuators

This group of EAP actuators are electronic, and use Maxwell pressure and capacitor structure to create unidirectional contractions. Stacked DE actuators consist of multiple layers of thin elastomers (~ 40 μm) coated with mechanically compliant electrodes, that is, multiple layers of compliant capacitors. When DC voltage is applied to this capacitor, Maxwell pressure compresses the thin DE in-between, and thereby the whole

Table 8.5 Longitudinal contraction of stacked DE actuator under Maxwell stress.

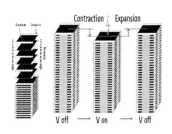

Advantages
High stress
Moderate strain
High strain-rate
Moderate efficiency
Commercially available
Length self-sensing
Disadvantages
High driving voltage

Stacked DE actuators consist of multiple layers of compliant capacitors. The advantages and disadvantages of these actuators are summarized in the right-hand column.
Source: Used with permission from Gabor Kovacs.

actuator contracts longitudinally (Table 8.5). The reported maximum force and unidirectional strain value for a cylindrical actuator of 20 mm of diameter and 25 mm length under 4.2 kV driving voltage were 32 N (102 kPa), and 18%, respectively [23].

Among the investigated actuators, only stacked DEA actuators were commercially available. Kovacs et al. tested the cycle life of these stacked DEA actuators at 500 cycles per day for 1 year. The result was a life of 182k cycles, with no observed fatigue. The reported strain-rate for the commercially available version of this actuator was roughly 500%/s [24]; and the typical reported efficiency for different DE actuators, was 25%–30% [5,12]. The DE actuators have energy recuperation capability; by converting their mechanical motion to electricity they can act as generators, which increases the efficiency, substantially [27]. In addition, most DE actuator configurations are able to self-sense their length. This property may be exploited to implement position control and monitoring systems without additional sensing components, thus reducing the mechanical complexity [28,29]. Unfortunately, the required driving voltage of 1 kV or higher [3,4] limits practicality as the available electronic components that can handle such high voltage are limited. The advantages and disadvantages of these actuators are summarized in Table 8.5.

8.3.5 Hydraulically amplified self-healing electrostatic actuators

These actuators combine the versatility of soft fluid actuators with the muscle-like properties of DE actuators. HASEL actuators are capacitors with a liquid dielectric and utilize Maxwell pressure to generate force. As applied voltage to the electrodes increases, Maxwell pressure increases; after reaching a specific threshold the electrodes abruptly approximate and displace the liquid dielectric under the electrodes to the adjacent surrounding volume. This is known as the snap-through transition and results in substantial hydraulic force (Table 8.6). Note that the use of liquid elastomers allows the maximum driving voltage to be increased to up to 40 kV without permanent electrical breakdown (also known as dielectric breakdown, that is, current flow through an electrical insulator after reaching a specific voltage) [22]. While resulting in very high stress and strain generation, this high driving voltage is more than 10 times that of the aforementioned stacked DE actuators, which substantially limits the applicability in wearable applications.

$$P = \varepsilon_0 \varepsilon_r \left(\frac{V}{d}\right)^2 \tag{8.2}$$

The higher the voltage, however, the higher the Maxwell pressure (Eq. 8.1), and thereby the higher the stress and strain generation. Under 14.5 kV driving voltage and actuation frequency of 2.4 Hz, a two-unit uniplanar HASEL actuator lifted 7 N of tensile load for a maximum stress of 114 kPa, 124% strain (roughly 600%/s strain rate) and 164 W/kg peak power-to-mass ratio. For a donut-like configuration, efficiency was reported at 21% with maximum actuation frequency of 20 Hz [22]. Under 1 kg of tensile load and 3.8 Hz of actuation, the cycle life of a uniplanar single-unit HASEL actuator was 158k cycles [22].

Table 8.6 Pros, cons, and actuation mechanism of the HASEL actuator.

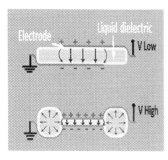

	Advantages
	High stress
	High strain
	High strain-rate
	Moderate power to mass ratio
	Moderate efficiency
	Length self-sensing
	Disadvantages
	Very high driving voltage
	Risk of leakage

When the voltage passes the snap-through threshold, the liquid abruptly moves into the surrounding area and creates substantial hydraulic force [22].

8.4 Discussion

The mechanical properties of five groups of soft actuators (CNF, EPC, PVC gel, stacked DE, and HASEL) were compared (Table 8.7). Stacked DEAs showed high stress-rate and moderate maximum strain and stress, which made this group of actuators a potential alternative for rehabilitation robotics. Their high driving voltage, however, limits the available electrical components for wearable applications.

Although the compared thermally driven actuators, that is, CNF and EPC, are powerful with high stress and unidirectional strain, their low strain-rates limit their applicability as artificial muscles. This is because actuator contraction time, which for skeletal muscle is very short (e.g., 36 and 12.5 ms for the soleus and digitorum longus muscles, respectively [27]), is limited by strain rate. Because the time delay between generating a command and applying the activation voltage to the actuator is usually very small in electrical systems, the contraction time is highly affected by the strain-rate. Therefore an actuator deployed for use in an assistive device requires fast strain-rates. Furthermore, the higher the actuation frequency the higher the controllability. Additionally, the low efficiencies of CNF and EPC actuators further limit their capabilities in wearable applications, where low energy consumption is vital. Although comparable to skeletal muscle in most measures, and superior to thermally driven actuators in efficiency and strain rate, EAP actuators (i.e., PVC gel, HASEL, and stacked DE) generated less unidirectional stress and strain than thermally driven actuators. For example, the efficiency for CNF, and EPC are 1.08% [7] and 0.2% [12], respectively, whereas stacked DEAs and HASEL actuators showed 30% [23] and 21% [22] efficiency, respectively.

Long cycle life is another major concern in active assistive devices, where many actuations are needed per day to assist a motion in activities of daily living. For example most major muscles that contributes during

Table 8.7 Summary of comparison between coiled nylon fiber (CNF), ethanol-based phase-change (EPC), poly vinyl chloride (PVC) gel, stacked dielectric elastomer (DE), and hydraulically amplified self-healing electrostatic (HASEL) actuators.

	CNF	EPC	PVC gel	Stacked DE	HASEL
Stress	High	High	Moderate	High	High
Strain	High	High	Moderate	Moderate	High
Strain-rate	Low	Low	Moderate	High	High
Efficiency	Low	Low	Moderate	Moderate	Moderate

walking gait activate at least once during a gait cycle [28], that is, two consecutive steps. Thus to assist a certain muscle for 5000 steps a day for a year, as an example, about 1 M cycle life is required. In our study, only CNF actuators (1.2 M cycles) achieved this [7]. The reported cycle life of the EAP actuators were 158k, 170k, and 182k cycles for HASEL, PVC, and stacked DE actuators, respectively [9,22,23].

Note that the reported cycle life for the stacked DE actuator was ambiguous in the literature. We assumed that the actuators underwent 182k cycles based upon our interpretation of the sentence "After 1 year in operation and approximately 500 repetitive activations no failure due to fatigue has been reported to date" [23]. Because the cycle life of the actuator was recently reported to be greater than 70 million cycles, by Dr. Kovacs for the commercially available version [24], we assumed the author meant 500 repetitive cycles per day for 1 year.

PVC gel actuators showed moderate stress and unidirectional strain, low power-to-mass ratio, 3 w/kg [9], and short cycle life (170k cycles [9]). Additionally, generating pushing force complicated the packaging design of these actuators (Fig. 8.3) and increased their bulk.

With high stress and strain-rate and moderate efficiency, and despite their short cycle lives of 158k and 182k cycles, respectively [22,23], HASEL and stacked DE actuators showed the most promise for rehabilitation applications. Furthermore, these actuators have a unique self-length-sensing feature, which is comparable to proprioception in biological muscle. This feature can decrease hardware complexity and manufacturing cost of the rehabilitation robots, through eliminating the need for additional displacement sensors. While HASEL actuators were superior to stacked DE actuators in unidirectional strain (to 124% [22] vs 18% [23], respectively), their higher driving voltage (up to 10 times), shorter cycle life, risk of leakage, and snap-through transition (an "all-or-none" type of actuation, which complicates feedback control strategies) make the stacked DE actuators the strongest candidate overall.

8.4.1 Chosen actuator

Stacked DE actuators were chosen due to moderate unidirectional strain (18%), high stress and strain-rates (102 kPa and >500%/s, respectively [23]), and—although graded as moderate—highest efficiency (25%−30% [5,12]). The commercial availability of CTsystems' stacked DE actuator (CT-SDEA) [24] was an additional advantage over the other investigated

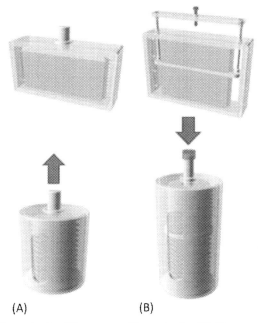

Figure 8.3 Pushing actuator (A) versus pulling actuator (B). The proposed packaging for the pushing PVC actuators is substantially bulkier than that of the pulling PVC gel actuators. *From Y. Li, M. Hashimoto, PVC gel based artificial muscles: characterizations and actuation modular constructions, Sens. Actuators A Phys. 233 (2015) 246–258.*

actuators in this study. These outstanding properties, in addition to softness, noiseless operation, and unidirectional contraction, make the stacked DE actuators suitable for forming artificial muscle in rehabilitation applications.

Note that although CT-SDEA is produced based on the aforementioned stacked DE actuators [23], the reported maximum force and unidirectional strain values are lower (Table 8.8). This difference is mainly because of the required long cycle life for commercial applications. To reduce the chance of electrical breakdown, CTsystems decreased the maximum activation voltage to 1.2 kV from the 4.2 kV used in Kovacs et al.'s study [23]. This increased cycle life to more than 70 M cycles, which is by far the longest of the compared actuators in this study, 1.2 M[7], 170k[9], and 158k cycles [22].

Table 8.8 Reported properties of the commercially available stacked DE actuator, CT-SDEA [24].

Reported properties in the manual	
Max force/stress	10 N/34.6 kN/m^2
Max strain	5%
Cycle life	70 M cycles
Actuation frequency	<100 Hz
Long-term stability	>5 years
Max driving voltage	1200 kV

8.5 Conclusion

Stress, unidirectional strain, power-to-mass ratio, strain-rate, and efficiency of five groups of soft actuators (CNF, EPC, PVC gel, stacked DE actuator, and HASEL) were compared. Stacked DE actuators were deemed most applicable as artificial muscle due to their high stress-rate, moderate maximum stress and strain, moderate efficiency, long life cycle, and commercial availability. The reliability of the reported electromechanical properties and additional measures, such as hysteresis and length tension curve, must be assessed for this commercially available group of actuators. These assessments are valuable for comprehending the limitations and capabilities of stacked DEAs in wearable robotics.

References

[1] D. Rus, M.T. Tolley, Design, fabrication and control of soft robots, Nature 521 (7553) (2015) 467.
[2] I.S. Yoo, S. Reitelshöfer, M. Landgraf, J. Franke, Artificial muscles, made of dielectric elastomer actuators - a promising solution for inherently compliant future robots, Soft Robot, Springer, 2015, pp. 33–41.
[3] Y. Bar-Cohen, V.F. Cardoso, C. Ribeiro, S. Lanceros-Méndez, Electroactive polymers as actuators, Advanced Piezoelectric Materials, *second ed.*, Elsevier, 2017, pp. 319–352.
[4] F. Carpi, R. Kornbluh, P. Sommer-Larsen, G. Alici, Electroactive polymer actuators as artificial muscles: are they ready for bioinspired applications? Bioinspir Biomim. 6 (4) (2011) 45006.
[5] J.D.W. Madden, N.A. Vandesteeg, P.A. Anquetil, et al., Artificial muscle technology: physical principles and naval prospects, IEEE J. Ocean. Eng. 29 (3) (2004) 706–728.
[6] T. Mirfakhrai, J.D.W. Madden, R.H. Baughman, Polymer artificial muscles, Mater. Today 10 (4) (2007) 30–38.

[7] C.S. Haines, M.D. Lima, N. Li, et al., Artificial muscles from fishing line and sewing thread, Science. 343 (6173) (2014) 868−872. Available from: 10.1126/science.1246906.
[8] M.D. Lima, N. Li, M. Jung de Andrade, et al., Electrically, chemically, and photonically powered torsional and tensile actuation of hybrid carbon nanotube yarn muscles, Science 338 (6109) (2012) 928−932. Available from: 10.1126/science.1226762.
[9] Y. Li, M. Hashimoto, PVC gel based artificial muscles: characterizations and actuation modular constructions, Sens. Actuators A Phys. 233 (2015) 246−258.
[10] J. Mohd Jani, M. Leary, A. Subic, M.A. Gibson, A review of shape memory alloy research, applications and opportunities, Mater. Des. 56 (2014) 1078−1113. Available from: 10.1016/J.MATDES.2013.11.084.
[11] J. Leng, X. Lan, Y. Liu, S. Du, Shape-memory polymers and their composites: stimulus methods and applications, Prog. Mater. Sci. 56 (7) (2011) 1077−1135.
[12] A. Miriyev, K. Stack, H. Lipson, Soft material for soft actuators, Nat. Commun. 8 (1) (2017) 596.
[13] D. Iqbal, M. Samiullah, D. Iqbal, M.H. Samiullah, Photo-responsive shape-memory and shape-changing liquid-crystal polymer networks, Mater. (Basel) 6 (1) (2013) 116−142. Available from: 10.3390/ma6010116.
[14] M. Kohl, D. Brugger, M. Ohtsuka, T. Takagi, A novel actuation mechanism on the basis of ferromagnetic SMA thin films, Sens. Actuators A Phys. 114 (2-3) (2004) 445−450. Available from: 10.1016/J.SNA.2003.11.006.
[15] Tadokoro Kim, K.J. Satoshi, Electroactive Polymers for Robotic Applications: Artificial Muscles and Sensors, Springer, London, 2007.
[16] Y. Li, M. Hashimoto, Design and prototyping of a novel lightweight walking assist wear using PVC gel soft actuators, Sens. Actuators A Phys. 239 (2016) 26−44.
[17] F. Carpi, S. Bauer, D. De Rossi, Materials science. Stretching dielectric elastomer performance, Science 330 (6012) (2010) 1759−1761. Available from: 10.1126/science.1194773.
[18] F. Carpi, D. De Rossi, R. Kornbluh, R.E. Pelrine, P. Sommer-Larsen, Dielectric Elastomers as Electromechanical Transducers: Fundamentals, Materials, Devices, Models and Applications of an Emerging Electroactive Polymer Technology, Elsevier, 2011.
[19] E. Biddiss, T. Chau, Dielectric elastomers as actuators for upper limb prosthetics: challenges and opportunities, Med. Eng. Phys. 30 (4) (2008) 403−418. S1350-4533 (07)00104-X.
[20] Q. Pei, R. Pelrine, S. Stanford, et al., Multifunctional electroelastomer rolls and their application for biomimetic walking robots, Synth. Metals. Int. Soc. Opt. Photonics (2003) 246−253. Available from: 10.1016/S0379-6779(02)00535-0.
[21] S. Ashley, Artificial muscles, Sci. Am. 289 (4) (2003) 52−59.
[22] E. Acome, S.K. Mitchell, T.G. Morrissey, et al., Hydraulically amplified self-healing electrostatic actuators with muscle-like performance, Science 359 (6371) (2018) 61−65.
[23] G. Kovacs, L. Düring, S. Michel, G. Terrasi, Stacked dielectric elastomer actuator for tensile force transmission, Sens. Actuators A Phys. 155 (2) (2009) 299−307.
[24] L.D. Gabor Kovacs, CTsystems, Swiss compliant transducer. <http://www.ct-systems.ch/>, 2016.
[25] M. de Onis, A.W. Onyango, WHO child growth standards, Lancet (London, Engl.) 371 (9608) (2008) 202−204. Available from: 10.1016/S0140-6736(08)60131-2.
[26] D.A. Winter, Biomechanics and Motor Control of Human Movement, John Wiley & Sons, 2009.
[27] R.I. Close, Dynamic properties of mammalian skeletal muscles, Physiol. Rev. 52 (1) (1972) 129−197. Available from: 10.1152/physrev.1972.52.1.129.

[28] M.W. Whittle, D. Levine, J. Richards, M.W. Whittle, Whittle's Gait Analysis, Elsevier Health Sciences, 2012.
[29] Hoffstadt, Maas, Model-based self-sensing algorithm for dielectric elastomer transducers based on an extended Kalman filter, Mechatronics 50 (2018) 248−258.

Index

Note: Page numbers followed by "*f*" and "*t*" refer to figures and tables, respectively.

A

Activation mechanism, 242
Active exoskeleton devices, 165–166
Active sensing, 43–44
Actuation, 90–91
   using Maxwell pressure, 244–245
   principle of XoSoft, 172–175
   technologies for physical human–robot interaction, 2–7
      high torque density motors, 4–5
      quasi-direct drive actuation, 5–7
Actuator with Adjustable Stiffness (AwAS), 201–224. *See also* Treadmill with Adjustable Stiffness (TwAS)
   antagonistic configurations, 201–209
      bidirectional antagonistic, 209, 209*f*
      cross coupling antagonistic, 208–209, 208*f*
      simple antagonistic, 202–208, 202*f*
   AwAS-II, 214–215, 214*f*
   configuration, 211–215, 212*f*
   drawbacks, 210–211
   experimental results on, 215–218
      preliminary experiments, 215–218
   mechanism, 212–213
   physical covered setup, 212*f*
   series configurations, 209–210, 210*f*
      based on pretension, 209–210
Actuator with mechanically adjustable series compliance (AMASC), 204–206, 204*f*
   cable routing diagram, 205*f*
   simple linear and rotational model, 205*f*
Actuators, need for, 91
Additive manufacturing (AM), 40–43
AFM, 55
Air, for actuators, 105
ALEX II worn exoskeleton, 136, 137*f*
Alpha version, XoSoft prototype, 167–168
Ammonium hydroxide (NH_4OH), 47–48
Ammonium peroxodisulfate (($NH_4)_2S_2O_8$), 45
Aniline ($C_6H_5NH_2$), 45
Anxiety (ANX), 193
Aquivion, 94
Artificial muscle applications, 104
Artificial skeletal muscle, 242
   actuation using Maxwell pressure, 244–245
   method, 245–246
      comparison measures, 245–246
      inclusion criteria, 245
   results, 246–252
   softness in comparison to biological tissue, 241*f*
Assistive devices, 165
Attitude toward using the technology (ATUT), 193
Augmented reality stepping task study, 155

B

Back exoskeleton, 26–33. *See also* Hip exoskeleton; Knee exoskeleton
   control architecture, 30–32
   design, 26–27
   evaluation, 32–33
   modeling, 27–30
Back injuries, 26
Backdrive torque, 13, 22
Behavioral Intention (BI), 193
Beta 1 and 2 versions, XoSoft prototypes, 168
Bidirectional antagonistic configuration, 209, 209*f*
Bidirectional Bowden cable mechanism, 15
Bilateral stiffness adjustment, 228–229, 229*f*
Bioinspired robotic jellyfish, 40–43
Biological inspired joint stiffness control, 202–203, 203*f*

Biomimetic(s), 91
  actuator, 111–112
  applications, 39–43
Biomimicry, 91
Biot–Savart law, 113–114, 128
Bipedal movement, 220
Braided artificial pneumatic muscle (BAPM), 105
Braided pneumatic actuators, 98–99
Braided shell of fibers, 105
Brushless direct current motors (BLDC motors), 3–5, 4f
Bucky gel actuator (BGA), 40–43

C
Cable attachment points, 140–141
Cable routing points, 140–141
Cable-driven Active Leg Exoskeleton (C-ALEX), 143–144, 144f, 150f, 151f
  for gait rehabilitation, 149–155
    augmented reality stepping task study, 155
    stroke rehabilitation case study, 149–155, 151f
Cable-driven Arm EXoskeleton (CAREX), 136–137, 138f
Cable-driven robots, 139
Cable-driven systems, 135–139
  categories, 142
  formulation of cable-driven devices, 139–141
  perturbation study using RobUST, 156–158
  for postural and gait rehabilitation with variable controllers, 142–149
Capacitive soft sensor, 179, 179f
Carbon (C), 69–71
  electrodes, 94–95
Catheters, 103
Center of mass (COM), 146
Center of pressure (COP), 152f, 159
Central nervous system (CNS), 232
Central pattern generator (CPG), 43–44
Charge model, 115–116
Circle packing density, 120–121

Clinical testing, 169, 192–194. *See also* Laboratory testing
  Exoscore XoSoft Beta 2, 194
  experimental protocol, 193
  XoSoft Beta 2 *vs.* XoSoft Gamma, 193–194
Coil average radius, 120–121
Coiled nylon fiber actuators (CNF actuators), 242, 247–248
Coiled polymer actuator (CPA), 99–100, 100f
  used in rehabilitative devices for wrists, 106
Compliant structures, 90, 106
Control schema considerations, 141
Controller Area Network bus (CAN bus), 5
Conventional actuation, 3
Conventional geared actuation, 2
Conventional rehabilitation, 90
Cross coupling antagonistic configuration, 208–209, 208f
CTsystems' stacked DE actuator (CT-SDEA), 254–255
Cyclic voltammetry (CV), 47
Cylindrical IPMCs, 96

D
Data acquisition (DAQ), 60
  card, 31–32
Data-driven method, 12–13
Degree of sulfonation (DS), 49–50
90 degrees turn, 185
Dehydration process of Kraton/GO/Ag/Pani polymer composite-based soft actuators, 46
Device work-space, 140–141
Dibutyl adipate (DBA), 92
Dielectric breakdown, 251–252
Dielectric elastomer actuators (DE actuators), 243
Digital analogue card (DAC), 49–50
Digital weighing/load cell, 63
Donning/doffing, 185
Doping effects of IPMC actuators, 40–43
Double distilled water (DMW), 48
DSP microcontroller, 31

E

EDX, 69–71
Efficiency, 246
Effort Expectancy (EE), 193
Ekso, 165–166
Electrical stimulation, 242
Electroactive polymers (EAPs), 39–40, 242. *See also* Ionic electroactive polymers (Ionic EAPs)
  actuators, 90–93
  EAP-based soft actuators, 40–43
  PVC gels, 92–93
Electrodeless IPMCs, 95
Electroless plating, 49
Electromagnetic actuators, 111, 112*f*
Electromagnetic clutch, 173
Electromagnetically driven elastic actuator, 111–112
Electromechanical efficiency, 246
Electromyography sensors (EMG sensors), 169
Electronic EAPs, 243
Elliptical wire geometry, 128–129
End effector-based devices, 136–137
EPC actuators. *See* Ethanol-based phase-change actuators (EPC actuators)
Ethanol-based phase-change actuators (EPC actuators), 242, 249
1-ethyl-3-methylimidazolium tetrachloro aluminate (IL), 39–40, 47–48
Eutectic gallium–indium (EGaIn), 112
Exoscore XoSoft Beta 2, 194
Exoskeletal brace, 90
Exoskeletons, 2, 136, 166
Experiential perception (EP), 193

F

FeRiBa3 (point of contact planar table-top robot), 136–137, 138*f*
Figure-of-8-walk, 193
Finite element method (FEM), 111–112, 224
Flemion, 94
Flexible link manipulator using SPVA/IL/Pt IPMCs for robotics assembly, 80–82
Foot clearance, 188
Foot-flat (FF), 175
Force sensing resistors (FSR), 176
Four-discrete-position electromagnetic actuator, 111–112
Fourier-transform infrared spectroscopy (FTIR spectroscopy), 47, 53, 71–72
Fractional-slot type winding, 4–5

G

Gamma version, XoSoft prototype, 168
Gear ratio, 11
Granular jamming principle, 172–173
Ground reaction forces (GRF), 185, 200–201

H

Harness systems, 200
Harvard soft exosuit, 166
Heel-off (HO), 175
Heel-strike (HS), 175
High torque density motors, 4–5
High-performance soft wearable robots
  actuation technologies for physical human–robot interaction, 2–7
  applications to wearable robots, 7–33
Hip exoskeleton, 7–14, 8*f*. *See also* Back exoskeleton; Knee exoskeleton
  control system, 12–13
  design, 7–8
  evaluation, 13–14
  modeling, 8–12
Hip joint support using polyvinyl chloride gels, 101–102
Human–exoskeleton coupled dynamic model, 8, 9*f*
Hybrid actuators (HAs), 90–91, 97–101
  HASEL actuator, 97–98, 98*f*
  nylon-based coiled polymer actuator, 99–101
  soft pneumatic and hydraulic actuators, 98–99
Hydraulic actuators, 98–99
  used in rehabilitative devices for hands and arms, 104–105
Hydraulically amplified self-healing electrostatic actuator (HASEL actuator), 90–91, 97–98, 98*f*, 104, 244, 251–252
Hydrogels, 102–103

I

Impedance spectroscopy, 51–52
In situ oxidative polymerization of aniline, 45–46
Inclusion criteria, 245
Indego, 165–166
Inertial measurement unit (IMU), 8, 178
Insole pressure sensors, 176–177, 177f
Institutional Review Board (IRB), 233
Intermediate link, 212–213
Ion exchange capacity (IEC), 47, 63–68
Ionic electroactive polymers (Ionic EAPs), 90–91, 93–96
   ionic hydrogels, 93–94
   IPMCs, 94–96
Ionic gel/metal nanocomposite (IGMN), 40–43
Ionic hydrogels, 93–94
Ionic liquid gel (IGL), 40–43
Ionic polymer metal composite (IPMC), 39–40, 90–91, 94–96, 242–243
   applications, 103–104
   base soft actuator by different approaches, 44–50
   fabrication, 96
   IPMC-based artificial muscle linear actuator, 40–43
   literature survey, 40–44
   results, 50–79
   robotic system, 80–82
   soft actuators, 39–40

J

Joint torques, 136–137

K

Kevlar, 206–207
Kinematics
   analysis, 27
   of human walking on TwAS, 231–238
      method, 233
      procedures and instrumentation, 233–234
      results, 234–236
Knee exoskeleton, 14–26. *See also* Back exoskeleton; Hip exoskeleton
   control system, 19–20
   design, 15–16
   evaluation, 20–26
   modeling, 16–19
Knee extension, 17–19
Knee extensors, 23–24, 25t
Knee flexors, 23–24, 25t
Knee joint torque, 17
Kneeling, 14–15
Kraton, 45
Kraton/graphene oxide/Ag/polyaniline polymer composite-based soft actuators (Kraton/GO/Ag/Pani polymer composite-based soft actuators), 39–40, 45–47
   characterizations, 47, 50–63
      AFM, 55
      electrochemical properties, 57–60
      electromechanical properties, 60–63
      FT-IR, 53
      SEM, 54–55
      UV–visible studies, 55–57
      water update properties, 50–52
   chemical composition, 46f
   dehydration process, 46
   fabrication, 45–46
   materials, 45
   proton conductivity, 47
Kraton/graphene oxide/Ag/polyaniline polymer composite-based soft actuators (Kraton/GO/Ag/Pani polymer composite-based soft actuators)

L

L-test, 193
Laboratory testing, 169, 184–192
   experimental protocol, 185–189
      body regions for evaluation of perceived exertion/pressure, 186t
      Borg scale, 186t
   XoSoft Beta 1, 189–190
   XoSoft Beta 2, 190–191
   XoSoft Gamma, 192
Linear actuators, 111–112
Linear model, 204
Linear sweep voltmmetry (LSV), 47
Liquid elastomers, 251–252

LiteGait, 200–201
Locomotion mechanics, 199, 227
Lokomat, 165–166
LOPES (gait training robot), 142, 165–166

M
Magnetic core, 115–118
Magnetization, 116
Manipulability of system, 139
Maximum Lyapunov exponent, 232
Maxon EC90 commercial motor, 4–5
Maxwell pressure, 243, 245f
　actuation using, 244–245
McKibben actuators, 105
Metabolic cost, 229–231
Mid-stance (MS), 175
Mid-swing (MSW), 176
Multiple shape memory ionic polymer–metal composite (MSMIPMC), 40–43
Multiple-shape memory effect, 40–43
MyoSwiss, 166

N
Nafion, 94–95, 104
Neodymium iron boron (NdFeB), 4–5
Neural network regressor, 12–13
Nitric acid (HNO_3), 47–48
Nonlinear finite-impulse response (NFIR), 43–44
Nylon-based coiled polymer actuator, 99–101

O
Optimized soft actuator based on interaction between electromagnetic coil and permanent magnet, 111–112
　influence of solenoid section deformation on magnetic field and force, 128–131
　manufacturing aspects and limitations, 125–128
　solenoid geometry design optimization, 120–125
　solenoid magnetic field and force calculation, 113–119
Optimization of solenoid, 120–125
Output link, 213
Oxygen (O), 69–71

P
Parallel cable-driven devices, 142
ParvoMedics VO_2max metabolic cost measurement system, 200–201
Passive sensing, 43–44
Peak flexion, 188
Pediatric exoskeleton, 241
Perceived adaptiveness (PAD), 193
Perceived usefulness (PU), 193
Perfluorinated membrane, 94
Perturbation study using RobUST, 156–158
Phase-change actuators, 242
Physical intelligence, 2
"Physically intelligent" robots, 2
Platinum (Pt), 69–71
Pleated pneumatic artificial muscle (PPAM), 206–208, 207f
Pneumatic actuation, 1–2
Pneumatic soft clutch, 174–175, 175f
PneuNets actuators, 104–105
Point of contact end-effector-based system, 136, 136f
Poly(3, 4-ethylenedioxythiophene)-poly (styrenesulfonate) (PEDOT:PSS), 40–43
Poly(vinyl alcohol) (PVA), 47–48
Polydimethylsiloxane (PDMS), 111–112
Polymer-based actuators, 242
Polymeric composite material-based actuators, 39–40
Polyvinyl chloride gels (PVC gels), 90–93, 242–243
　actuators, 249–250
　hip joint support using, 101–102
Postural and gait rehabilitation with variable controllers, cable-driven systems for, 142–149
Posturography, 193
Power-to-mass ratios, 243, 246
Powered exoskeletal devices, 1
Primary actuation methods, 2, 3f
Primary users (PU), 167, 170–171

Index

Proportional control method of solenoid actuator, 111–112
Proportional-integral-derivative (PID) control system, 19, 49–50
Proprioceptive actuation. *See* Quasi-direct drive actuation (QDD actuation)
Proton conductivity (PC), 39–40, 63–68
  of Kraton/GO/Ag/Pani polymer composite-based soft actuators, 47

Q

Quality-of-life enhancement (QoLE), 193
Quasi-direct drive actuation (QDD actuation), 2–3, 5–7
Quasi-static model, 17, 17f
Quasipassive actuation, 172

R

Range of motion, 188
Range of stiffness adjustment, 225–227
Reference assistive force, 30–31
Rehabilitation, 89–90, 165
  robotics, 135–136
ReWalk, 165–166
Rigid-link exoskeletons, 136
Robotic. *See also* Soft robotic(s)
  assembly, 39–40
  assistance in rehabilitation, 90, 135–136
  fish, 40–44
  orthosis, 142
  system, 80–82
    flexible link manipulator using SPVA/IL/Pt IPMCs for robotics assembly, 80–82
Robotic Upright Stand Trainer (RobUST), 146–148, 147f
  perturbation study using, 156–158
Romberg test, 193
Rotational model, 204
Routine diagnostic instrumented gait analysis, 193

S

Scanning electron microscopy (SEM), 47, 49–50, 54–55, 69–71

Secondary users (SU), 167, 171
Self-efficacy (SE), 193
Self-liberty (SL), 193
Self-sensing actuation (SSA), 43–44
Sensing methods, 43–44
Serial cable-driven devices, 142
Series elastic actuation (SEA), 2–3
Shape memory, 95
Shape memory alloys (SMAs), 242
Shape memory polymers (SMPs), 242
Simple antagonistic configuration, 202–208, 202f
Simulated home environment validation, 194–195
Slit IPMC cylindrical/tubular elements, 43–44
Snap-through transition, 251–252
Social influence (SI), 193
Sodium borohydride ($NaBH_4$), 47–48
Sodium nitrate ($NaNO_3$), 47–48
Soft actuation, 179–180
Soft actuators, 39–40, 241
Soft pneumatic actuators, 98–99
  used in rehabilitative devices for hands and arms, 104–105
Soft robotic(s), 90
  actuators, 91–101
    EAPs, 91–93
    hybrid actuators, 97–101
    ionic EAPs, 93–96
  applications, 101–106
    coiled polymer actuators used in rehabilitative devices for wrists, 106
    HASEL actuators and artificial muscle applications, 104
    hip joint support using polyvinyl chloride gels, 101–102
    hydrogels, 102–103
    ionic polymer–metal composite applications, 103–104
    soft pneumatic and hydraulic actuators used in rehabilitative devices, 104–105
Soft sensors, 178–179
Solenoid's wire length, 120
Spinal cord injury (SCI), 165

Spine-inspired continuum soft exoskeleton, 26
Spring force response, 206, 207f
Squat assistance control strategy, 20—21, 21f
Squatting, 14—15
Stacked DE actuators, 244
Stacked dielectric elastomer actuators, 250—251, 254—255
Standard deviation (SD), 63—68
State-space dynamic model, 43—44
Step length asymmetry, 188
Stiffness motor, 212—213
Straight walking, 185
Strain-limiting layers, 98, 104—105
Strain-rate, 246
Stress, 246
Stride cycle, 234
Stride interval dynamics, 232
Stride length, 188
Stroke rehabilitation case study, 149—155, 151f
Sulfonated graphene oxide, 40—43
Sulfonated poly(1, 4-phenylene ether-ether-sulfone), 40—43
Sulfonated polyvinyl alcohol (SPVA), 39—40
   SPVA/IL/Pt IPMC composite-based IPMC soft actuator, 47—50
      characterizations, 49—50, 63—79
      electrical properties, 72—73
      electromechanical properties, 73—79
      fabrication, 49
      materials, 47—50
      preparation of reagent solutions, 48
      preparation of SPVA membrane, 48
4-sulfophthalic acid, 47—48
Sulfur (S), 69—71
Surface charge density, 117
Surface electromyography (sEMG), 185
Surface stiffness, 199
   effects on human gait and metabolic cost, 229—231
   independence of, 227—228
System usability scale (SUS), 186—188, 187t

T
TadRob, 40—43
Task-oriented training, 89—90
Tertiary users (TU), 167
Tethered Pelvic Assist Device (TPAD), 148—149, 148f
Tetrahydrofuran (THF), 45
Textile jamming, 172—173, 173f
Thermally activated actuators, 242
Timed up and go test (TUG test), 232
Toe-off (TO), 176
Treadmill with Adjustable Stiffness (TwAS), 200—201, 219—224
   design specification, 221—224
   experimental results for stiffness adjustment of, 224—238
   force transmission module, 222f
   kinematics of human walking on, 231—238
   performance specification, 221
   stiffness adjustment mechanism, 201, 223f, 224f
Trunk Support Trainer (TruST), 145—146, 145f
Trust, 193
Turn density, 122
Two-dimensional motion dynamics (2D motion dynamics), 43—44

U
Unidirectional strain, 246
User-centered design approach (UCD approach), 166
   XoSoft EU Project, 166—169, 168f
UV—visible studies, 55—57

V
Validation in home-simulated environments, 169
Variability in feedback, 155
Variable stiffness, 201, 212—213
Vicon cameras, 200—201
Virtual impedance model, 33

W
Waist frame, 7—8
Walkways, 199

Water loss, 68–69
Water update properties, 50–52
Water uptake (WU), 39–40
Water-holding capacity (WH capacity), 49–50, 63–68
Wearable robots, 1–2
  applications to, 7–33
  emulator, 15
Whipping effect, 191
Wireless tadpole robot, 40–43

X

X-ray diffraction (XRD), 49–50, 71–72
XoSoft Beta 1 prototype, 179–180, 180f
  laboratory testing, 189–190
XoSoft Beta 2 prototype, 180–181, 181f
  clinical testing, 193–194
  laboratory testing, 190–191
XoSoft EU Project, 166
  actuation principle, 172–175
  prototypes, 179–183, 183t
    comparison, 182–183
    XoSoft Beta 1 prototype, 179–180, 180f
    XoSoft Beta 2 prototype, 180–181, 181f
    XoSoft Gamma prototype, 181–182, 182f
  requirements, 169–171
  sensing and control, 175–179
  testing and validation, 184–195, 184t
  UCD approach, 166–169, 168f
XoSoft Gamma prototype, 181–182, 182f
  clinical testing, 193–194
  laboratory testing, 192

Y

Young's modulus, 242

Printed in the United States
By Bookmasters